Climate Variability and Water Dependent Sectors

Demand for water in agricultural, municipal, industrial, and environmental uses is growing. More frequent and severe extreme weather conditions now exacerbate water shortages in many locations and existing infrastructure to store and release water rarely has the capacity to both prevent floods during wet periods *and* meet demand during drought periods. Competition among sectors adds pressure not only on water infrastructure, but also on management policies and allocation institutions.

This book of contributed chapters assesses the performance of existing infrastructure, institutions and policies under different climate variability scenarios. It also provides suggestions for minimizing conflict over scarce water resources. More flexible water-allocation institutions and management policies, and better tools for decision-making under uncertainty will be required to maximize society's net benefit from less reliable water resources. The chapters show how incentives for individuals to conserve water, and policies for helping vulnerable populations prepare for and recover from extreme events, will also need to be improved.

This book was originally published as a special issue of the *Journal of Natural Resources Policy Research*.

Dannele E. Peck is an associate professor in the Department of Agricultural and Applied Economics at the University of Wyoming, USA.

Jeffrey M. Peterson is a professor in the Department of Agricultural Economics at Kansas State University, USA.

Climate Variability and Water Dependent Sectors

Impacts and Potential Adaptations

Edited by
Dannele E. Peck and Jeffrey M. Peterson

Routledge
Taylor & Francis Group

LONDON AND NEW YORK

First published 2015 by Routledge

2 Park Square, Milton Park, Abingdon, Oxfordshire OX14 4RN
711 Third Avenue, New York, NY 10017

Routledge is an imprint of the Taylor & Francis Group, an informa business

First issued in paperback 2018

British Library Cataloguing in Publication Data
A catalogue record for this book is available from the British Library

ISBN 13: 978-1-138-80733-4 (hbk)
ISBN 13: 978-1-138-38384-5 (pbk)

Typeset in Times New Roman
by RefineCatch Limited, Bungay, Suffolk

Publisher's Note
The publisher accepts responsibility for any inconsistencies that may have
arisen during the conversion of this book from journal articles to book chapters,
namely the possible inclusion of journal terminology.

Disclaimer
Every effort has been made to contact copyright holders for their permission to
reprint material in this book. The publishers would be grateful to hear from any
copyright holder who is not here acknowledged and will undertake to rectify
any errors or omissions in future editions of this book.

Contents

Citation Information

The chapters in this book were originally published in the *Journal of Natural Resources Policy Research*, volume 5, issue 2–3 (April–July 2013). When citing this material, please use the original page numbering for each article, as follows:

Chapter 1
Introduction to the special issue on 'Climate Variability and Water-Dependent Sectors: Impacts and Potential Adaptations'
Dannele E. Peck and Jeffrey M. Peterson
Journal of Natural Resources Policy Research, volume 5, issue 2–3 (April–July 2013)
pp. 73–78

Chapter 2
Cost of early snowmelt in terms of reduced irrigation deliveries
Aaron Benson and Ryan Williams
Journal of Natural Resources Policy Research, volume 5, issue 2–3 (April–July 2013)
pp. 79–90

Chapter 3
Climate change opportunities for Idaho's irrigation supply and deliveries
Russell J. Qualls, R. Garth Taylor, Joel Hamilton and Ayodeji B. Arogundade
Journal of Natural Resources Policy Research, volume 5, issue 2–3 (April–July 2013)
pp. 91–106

Chapter 4
Climate and choice of irrigation technology: implications for climate adaptation
George B. Frisvold and Shailaja Deva
Journal of Natural Resources Policy Research, volume 5, issue 2–3 (April–July 2013)
pp. 107–128

Chapter 5
Potential economic impacts of water-use changes in Southwest Kansas
Bill Golden and Jeff Johnson
Journal of Natural Resources Policy Research, volume 5, issue 2–3 (April–July 2013)
pp. 129–146

Chapter 6
Community adaptation to climate change: exploring drought and poverty traps in Gituamba location, Kenya
Amy Sherwood
Journal of Natural Resources Policy Research, volume 5, issue 2–3 (April–July 2013)
pp. 147–162

Chapter 7
Medium-term electricity load forecasting and climate change in arid cities
Bhagyam Chandrasekharan and Bonnie Colby
Journal of Natural Resources Policy Research, volume 5, issue 2–3 (April–July 2013)
pp. 163–182

Chapter 8
The joint impact of drought conditions and media coverage on the Colorado rafting industry
Karina Schoengold, Prabhakar Shrestha and Mark Eiswerth
Journal of Natural Resources Policy Research, volume 5, issue 2–3 (April–July 2013)
pp. 183–198

Please direct any queries you may have about the citations to
clsuk.permissions@cengage.com

Introduction

Dannele E. Peck[a] and Jeffrey M. Peterson[b]

[a]Department of Agricultural & Applied Economics, University of Wyoming; [b]Department of Agricultural Economics, Kansas State University

Water stress is an increasingly important challenge for the twenty-first century. Demand for water resources in agricultural, municipal, industrial, and environmental uses is rising due to population and income growth. Climate change is expected to exacerbate water stress by increasing the frequency and severity of extreme weather conditions. Although drought may not become more frequent or severe in all locations, growing demand for water will cause scarcity to become more pressing over larger geographic areas. In many locations, infrastructure to store and release water currently lacks the capacity to meet demand during drought periods and prevent floods during wet periods. Competition for scarce water resources, both within and among sectors, will place even greater pressure on existing infrastructure, as well as current water allocation institutions and management policies.

Reassessment of infrastructure capacity and management, including the timing of storage releases, will be necessary to minimize conflicts between competing uses. More flexible water-allocation institutions and management policies, as well as better tools for decision-making under uncertainty, will be required to maximize society's net benefit from scarce water. Effectiveness of incentives for individuals to conserve water, and policies for helping vulnerable populations and sectors prepare for and recover from extreme events, will also need to be improved.

This special issue contains seven research articles, which are purposely ordered to follow water's path through the sweep of the hydrologic cycle affecting human decisions and livelihoods: from its origin as snowpack in distant headwaters to its capture in multi-purpose reservoirs, and its release for a diverse set of consumptive and non-consumptive uses by millions of downstream producers and consumers. Most of the contributed articles focus on the western United States, which is similar to a number of semi-arid regions in the world where water largely originates as mountain snowpack. Spring and summer snowmelt slowly seeps into ephemeral streams, which converge into roaring rivers that ultimately recharge aquifers and fill an intricate system of natural and constructed reservoirs. Society faces a complex set of decisions to manage the water in these storage basins, requiring difficult tradeoffs in allocating a scarce supply across the demands of numerous stakeholders.

Depending on its timing and intensity, water can be a productive asset or a destructive natural force. Constructed reservoirs help dampen the effects of too much water or too little water, along with the effects of water arriving too early or too late. It can be difficult, however, to manage reservoirs in a manner that achieves socially desirable outcomes, especially when multiple users have competing objectives.

Benson and Williams' opening contribution explores the circumstances in which flood-protection and irrigation water supply might be competing objectives. They model reservoir inflows and releases under historical versus future climate conditions. Climate change is expected to alter the timing of snowmelt, causing inflows to arrive earlier in the spring and over a shorter duration. To ensure that peak inflows will not lead to flood damage to downstream communities, reservoir managers must leave more excess storage capacity early in the runoff season. The authors show that the result can be far less runoff to be captured overall, leading to lower deliveries to farmers in the subsequent irrigation season. They also show that the tradeoff between flood protection and irrigation depends, in part, on a reservoir's storage capacity relative to downstream agriculture's water demand.

Qualls, Taylor, Hamilton and Arogundade continue the discussion of reservoir management under future climate conditions, but examine the strengths and weaknesses of an existing allocation mechanism. In the western US, the most prevalent legal institution to allocate water is the prior appropriation doctrine, which, in theory, concentrates water shortages on users holding the most junior water rights. In practice, water use is limited by the availability in a given location, and even senior users accessing water from a low-capacity reservoir may suffer shortages in times of drought. The authors find that although some users are likely to suffer occasional shortages in future drought years, the average shortages are relatively small. However, above-average precipitation might also occur under future climate conditions, and the authors' results also suggest that the current system is unable to store and allocate these above-average water supplies. Climate change could actually benefit water users in Idaho, but only if reservoir infrastructure management adapts appropriately.

Upon its release from natural or constructed storage basins, water's most likely user is irrigated agriculture, which consumes more water than any other user-group in the western US. Irrigated crop producers must be especially vigilant and responsive to changing climate because of their economic dependence on water quantity and timing, and other climatic variables during the growing season. As conditions evolve, so too will producers' production practices.

Frisvold and Deva explore the effects of temperature, growing season length, and irrigation costs on agricultural producers' adoption of sprinkler versus gravity-fed irrigation. Sprinkler irrigation may be a useful adaptation to changing climate conditions in some, but not all, regions of the western US. In many cooler regions, climate change is expected to increase the number of frost-free days while also raising the risk of frost during the growing season. In these locations, sprinkler irrigation may be beneficial to protect crops from early-season frost damage. In regions that already have high temperatures and long growing seasons, such as California and Arizona, sprinkler irrigation may not be a beneficial adaptation because warming conditions would increase evaporative losses during water delivery. The authors' regression analysis of past adoption in different states confirms that climatic variables, as well as farm size and farm input prices, are important determinants of sprinkler adoption rates.

Golden and Johnson continue the assessment of irrigated agriculture's vulnerability and adaptation to water shortages (current or future), but draw our attention to the Ogallala Aquifer in southwest Kansas, where decades of sprinkler-irrigated agriculture have significantly depleted the aquifer. In some areas, groundwater depletion is so severe that producers now face the difficult decision of whether to exhaust the aquifer, or reduce current consumption in an effort to extend its life for future generations. The authors estimate the potential benefits and costs of groundwater conservation, from the perspective of individual producers and the regional economies of three agriculture-dependent communities. Results suggest that groundwater conservation has the potential to generate net benefits for certain producers and regional economies, especially under drought conditions. For some regions, however, the short-term costs of groundwater conservation might exceed the long-term benefits. Conservation policies should therefore be tailored for each community's unique circumstances.

Individual producers in the US have access to numerous drought preparedness and mitigation tools, including private or government-sponsored safety nets, such as crop insurance and disaster assistance. In developing countries, however, these formal safety nets rarely exist, and individual producers have limited access to physical or financial resources necessary to adequately prepare for or mitigate the effects of drought.

Sherwood takes us to central Kenya, where a majority of households still engage directly in agricultural production. Few households can afford to drill groundwater wells or install advanced irrigation systems. Hand-watering is more common, but requires more labor than many households can afford to divert from other critical sources of income, such as off-farm employment. Drought is therefore more likely to lead to crop failure and life-threatening food insecurity. Sherwood uses participatory research methods to identify the consequences of a severe water shortage for women in a rural agricultural community, as well as their sources of vulnerability and resiliency. Agricultural households' preparedness for future droughts could be improved by reducing sources of vulnerability that erode their ability to build private or community-based safety nets during non-drought years. Despite being thousands of miles from the western US, these insights are relevant to any agricultural household whose livelihood depends on their ability to make the most of good growing conditions, and minimize losses during difficult times.

Few households in the US engage directly in agricultural production. These non-agrarian households tend to be more insulated from drought's negative effects on food production. They also have access to more resources for adapting to and mitigating the effects of drought and other features of climate change. Nonetheless, US households are not immune to the effects of water shortages. They rely, after all, on goods and services provided by water-dependent industries, such as agriculture, energy, and outdoor recreation. By understanding the effects of climate variables on consumer behavior and resulting demand, water-dependent businesses can minimize more-effectively the negative economic impacts of drought and climate change.

Chandrasekharan and Colby isolate the effects of weather conditions on consumers' use of electricity in a major metropolitan area of Arizona. They estimate the economic value to electric utilities of better-understanding and anticipating consumers' response to future climate conditions, especially changes in peak load during summer months when water for energy production is most scarce. Improved electricity load forecasts, based on medium-term weather expectations, should reduce electric utilities' cost of over or under-production.

Lastly, *Eiswerth, Schoengold and Shrestha* explore the sensitivity of commercial white-water rafting customers to drought, air temperature, and wildfire. After controlling for the effects of drought and other relevant physical and economic variables, customer numbers along the Arkansas River in Colorado were lower than expected in 2002, a year in which wildfires ravaged the general region. The authors hypothesize that imprecise national media coverage of wildfires in Colorado had unintended negative consequences for recreation-dependent businesses in the region, even though the safety and quality of rafting in the study area was not adversely affected.

While tracing water's journey from its origin as mountain snowpack to its capture in reservoirs and aquifers, and its use in agriculture, electricity and outdoor recreation, the seven articles in this issue highlight the diverse, situation-specific effects of climate variability on different sectors. As the varied research problems require, these seven studies apply different methods and obtain distinct results. Common themes nevertheless emerge from this varied work, four of which we discuss below.

(1) Water's increasing scarcity and variability generate difficult tradeoffs that require individuals, and society as a whole, to make complex decisions, often under imperfect information. Reservoir managers must balance the needs of flood protection, agriculture, recreation and other potentially competing uses when planning the quantity and timing of reservoir releases. Agricultural producers must weigh short versus long-term tradeoffs when choosing irrigation technology and water application rates. Producers in the US have more options and resources available to them than smallholder farmers in Kenya, but both should anticipate the potential for higher opportunity costs of acquiring and consuming water under future climate conditions, and begin assessing their existing sources of vulnerability and resiliency.

(2) Although climate change, in many cases, will heighten the scarcity of water resources and raise the stakes of already difficult tradeoffs, it will generate some new opportunities. These opportunities are potentially beneficial to society, if we position ourselves to take advantage of them. Reservoirs in some regions, or with certain physical characteristics, may be able to capture earlier, faster, and above-average inflows in some years and store them for beneficial use in below-average years. Groundwater conservation, in some cases, could generate long-term benefits, such as buffering against future droughts, that outweigh short-term costs. Commercial rafting companies could mitigate some of the effects of drought by marketing their services to families and inexperienced rafters who seek low-risk rafting experiences.

(3) To maximize society's net benefit from scarce water resources under future climate conditions, resource managers, businesses and policymakers need to understand how climate influences individual water-users' decisions and associated outcomes. This knowledge can help them prepare for, and adapt to, changing water availability and variability more effectively. Electric utilities, for example, can use medium-term weather information to improve peak load forecasts and reduce the cost of over or under-production. Outdoor recreation companies can coordinate with relevant media outlets to ensure potential customers are receiving accurate information about conditions in specific locales.

(4) Society's preparedness and adaptation activities should, ideally, be tailored to fit affected regions, communities, or water-user groups' unique characteristics, strengths and weaknesses. Local conditions will influence whether future climate conditions will exacerbate current water management challenges or create new opportunities, and whether proposed adaptation policies will increase or decrease society's net benefit from scarce water resources. A reservoir's resiliency to changing climate, for example, depends on its storage capacity relative to downstream demand and the value of properties within future floodplains. The appropriateness of adopting sprinkler irrigation or conserving groundwater

depends on individual farms and regions' physical and socio-economic characteristics, such as soil type, growing season length, and discount rate. For smallholder farmers in Kenya, the design of drought preparedness plans and climate change adaptation should be informed by existing sources of vulnerability and resiliency within the community and across households, including those that exist during non-drought years.

The seven articles in the pages that follow provide policy-relevant results for a number of sectors in the western US, which will be of direct interest to stakeholders and policymakers in the region. A number of the results and approaches are also applicable to numerous regions of the world that resemble the snowmelt-driven, semi-arid climates studied in this work. More broadly, these articles collectively bring out a set of tools and insights that can inform a range of management decisions as water-dependent and other resource sectors adapt to climate variability.

Acknowledgements

Coordination and editing of this special issue was supported by USDA-NIFA Multistate Research Project W-2190, 'Water Policy and Management Challenges in the West'.

Cost of early snowmelt in terms of reduced irrigation deliveries

Aaron Benson[a] and Ryan Williams

[a]*Agricultural and Applied Economics, Texas Tech University*

Climate change and windborne dust are expected to change snowmelt timing. Dust from agricultural activities is estimated to cause snow to melt two to four weeks earlier in some regions. Early snowmelt may decrease deliveries of irrigation water when reservoirs provide flood protection and irrigation water. To date, the quantity decrease has not been determined. We identify when early snowmelt causes earlier water release, and calculate the quantity decrease in water deliveries downstream. Irrigation reductions are negligible unless the dam capacity is less than twice the quantity of annual runoff, and unless snowmelt occurs more than four weeks early.

1. Introduction

Many irrigators and municipalities in western North America depend critically upon water that precipitates as snowfall and is temporarily stored as snowpack at high elevations. Reservoirs and water delivery systems in the region have been designed considering the timing and quantity of runoff as that snowpack melts. However, certain factors such as climate change and increased windborne dust and soot from human activities can cause snow to melt earlier than it would otherwise, or earlier than what was expected when the water delivery infrastructure was designed. In this paper, we investigate the economic effects of earlier runoff in the case of a small reservoir that is designed to provide flood control as well as water for irrigators downstream.

In the future, springtime snowmelt is expected to occur earlier in the year due to several factors, including climate change (Leung et al., 2004; Stewart, Cayan, & Dettinger, 2004;), and deposition of windborne dust and soot. Previous research has shown that most climate change scenarios increase the occurrence of extreme weather events and increase the variance of temperature and precipitation (Rowell, 2005; Seneviratne, Lüthi, Litschi, & Schär, 2006;), which can directly lead to snow melting earlier. Windborne dust and soot that settles on snowpack decreases the reflectivity of the snow, increasing the amount of solar radiation absorbed, and increasing the speed at which the snow melts (Painter et al., 2007; Qian, Gustafson, Leung, & Ghan, 2009). Much of the dust deposited on mountain snowpack in the western United States is from human activities, primarily cattle grazing (Neff et al., 2008). In simulations of snowmelt in the San Juan Mountains of southwest Colorado, windborne dust has been estimated to cause snow to melt 18 to 35 days earlier than in the absence of dust (Painter et al., 2007). Early snowmelt has been shown to have various biological and ecological effects (Steltzer, Landry, Painter, Anderson, & Ayres, 2009), and Stewart et al. (2004) recognized that, for reservoirs that provide flood protection in addition to other services, 'earlier flows would place more of the year's runoff into the category of hazard rather than resource' (p. 230), due to the facility needing to release more water earlier in the season than normal. The characteristics of snowmelt and storage facilities that might result in decreased irrigation deliveries later in the year were not identified, however, and the economic consequences of earlier snowmelt have, to our knowledge, not yet been determined.

Regardless of the cause of early snowmelt, it is necessary to first identify the cost of accelerated snowmelt runoff before designing policies to mitigate those costs. Once an estimate of costs of early snowmelt is available, it will be possible to identify the optimal investment in strategies (to adjust infrastructure, for example) to adapt to early snowmelt from climate change. Likewise, if early snowmelt is caused by dust deposition, an accurate estimate of costs is necessary to determine the optimal abatement of dust from ranchers and farmers, similar to a classic nonpoint source pollution problem. This scenario differs from many of the common examples of nonpoint source pollution, however, in that at least some of the agricultural producers will directly bear some of the 'externality' cost caused by windborne dust. This characteristic opens the door to policy instruments that are more common in provision of public goods, rather than just pollution reduction, since the reduction that results will benefit those who choose whether or not to participate in abatement. So, policies to reduce dust that causes early snowmelt will potentially need to include provisions to handle free-riders who allow others to adjust management to reduce dust and then benefit without bearing any abatement costs themselves. Again, however, damage costs need to be estimated before reduction policies can be considered.

To determine those costs, we create a simplified model of a small reservoir that stores spring runoff water for irrigation later in the year and also provides flood protection for a municipality downstream. We choose a small reservoir to simplify the model by avoiding the multi-year decisions that would be made by a manager of a large reservoir. Given the dual objectives of the dam, the dam manager must choose the quantity of water to release during spring runoff to both maximize the benefits of water stored for irrigation and minimize the expected monetary damages of releasing a flood-level quantity of water, which would be necessary if the reservoir is at capacity when a flood-level quantity of runoff enters the reservoir (a flood-level quantity of water is defined here as a flow rate greater than the capacity of the river/stream system below the dam).

Previous work has investigated the interaction of different objectives of reservoir management. Dudley (1988) examined the potentially conflicting short-, intermediate- and long-run irrigation objectives of a reservoir, when irrigation requirements and water availability are both stochastic. Vedula and Mujumdar (1992) and Vedula and Kumar (1996) considered reservoirs that provide only irrigation water, but that provide water for different crops with different water requirements, and built stochastic dynamic programming models to determine optimal water allocation to the different crops. Chatterjee, Howitt, and Sexton (1998) created a dynamic optimization model to determine the optimal allocation of water for irrigation and hydropower, recognizing that peak irrigation demand does not occur concurrently with peak hydropower demand, and find that actual water allocation in California deviates from the optimum. Lee, Yoon, and Shah (2011) modeled an entire watershed that includes a reservoir in order to determine optimal soil conservation upstream of the dam, reservoir sediment removal, water deliveries to downstream farmers and pollution control downstream of the dam, and find that an integrated watershed management plan can substantially increase the value of the watershed compared to a scenario in which the goals are considered individually.

In addition to modeling a dam with multiple, competing objectives, we also model a significant change in the characteristics of the runoff water the dam is meant to hold, which may require an adjustment from the dam manager, somewhat similar to Palmieri, Shah, and Dinar (2001) who studied the possibility of changing dam management strategies to address unsustainable sedimentation and increase the life of the reservoir. Other studies of streamflow timing and irrigation include Heidecke and Heckelei (2010), who investigated the effects on irrigation and farm income of decreasing water inflow to a single-use reservoir in Morocco, and found that the probability of the farmers receiving revenues below the subsistence level increases sharply under different climate change scenarios.

2. Model

We model the decisions made by a dam manager during a single runoff season to hold or release water, assuming that the releases after the runoff season are predetermined. We assume that the manager knows only the quantity of water currently held in the reservoir and the probability of a flood, which is determined by fraction of the reservoir's capacity currently filled and the number of days until the irrigation season begins. During runoff, the dam manager holds and releases water to satisfy the dual objectives of maximizing the benefits of water available for the later irrigation season and of minimizing the expected costs of having to release a flood-level quantity of water.[1] These two objectives will lead the manager to impounding water during spring runoff until the marginal benefit of stored water equals the marginal cost of using additional storage capacity. The marginal benefit of additional stored water is simply the additional revenue received by farmers when that water can be used to irrigate crops. The marginal cost of using additional storage capacity is the increase in the expected damage cost of having to release a flood quantity of water. Throughout this paper, we consider the effects of early snowmelt on irrigation deliveries in a year with average snowfall.

We assume D is the (constant) damage cost of a flood, and O is the probability of having to release a flood quantity of water, given t days until the irrigation season begins and ε percentage of the reservoir filled, so that the expected damage cost is $D^*O(t,\varepsilon)$, and the marginal cost of impounding additional water is $D^*\partial O(t,\varepsilon)/\partial\varepsilon$. The dam manager satisfies the dual objectives of the dam by choosing a threshold value of ε to maximize $B(\varepsilon) - D^*O(t,\varepsilon)$. As runoff flows into the reservoir on a given day, the dam manager will allow the water level to rise up to point where $MB = D^*\partial O/\partial\varepsilon$, or until $MB/D = \partial O/\partial\varepsilon$, that is, until the ratio of the value of one additional unit of water for irrigation to the damage costs of a flood just equals the increase in the probability of flood due to filling 1% of capacity. Once the optimal threshold is reached, the manager will immediately release additional inflows to keep the water level constant.

We assume that the probability of flooding, O, increases with ε (at an increasing rate) and that the marginal increase in probability of flooding, $\partial O/\partial\varepsilon$, decreases with t. These assumptions, along with an assumption of constant or diminishing marginal benefits of stored water, imply that the dam manager's threshold level increases as the season progresses. Depending on the timing of runoff, it is possible that this effect (of storing less water earlier in the season) could lead to less water available when the irrigation season begins under early runoff than under normal runoff.

The size of total storage capacity of the reservoir compared to annual irrigation obligations critically determines whether early snowmelt decreases irrigation deliveries in a single year. Consider a reservoir that has total storage capacity substantially greater than irrigation obligations, which is the case in our example referenced below. If there is a small amount of water stored when runoff begins, even if runoff occurs early, the optimal storage threshold would not be reached before irrigation started, and all of the runoff water could be used for later-season irrigation. If, instead, the reservoir was closer to capacity when runoff begins early, the manager would release much of the snowmelt runoff, but since the reservoir was close to capacity, which is greater than planned irrigation deliveries, irrigation obligations could be easily met.

If the total reservoir storage capacity is closer to total irrigation obligations, then early snowmelt would reach the optimal storage threshold early in the runoff season. Even if the quantity of water in the reservoir is relatively low when runoff begins, the manager would optimally hold a lower maximum quantity of water, meaning that something less than reservoir capacity would be held over for irrigation. If peak runoff occurs early in the runoff season, well before irrigation normally begins, and the manager is unable to hold much of that water, then it may not be possible to fulfill irrigation obligations in this case, depending on the relative size of those obligations to the runoff quantity. We assume that farmers are unable to adjust their growing seasons earlier to capture some of the released runoff water, although such an action may be possible if warmer temperatures allow farmers to anticipate early snowmelt. These adjustments would mitigate the costs of early snowmelt we estimate below.

In addition to the size of obligations relative to snowmelt quantity, early snowmelt may or may not result in reduced irrigation deliveries depending on the nature of the snowmelt and runoff. Reservoirs and watershed basins have different average runoff profiles that occur at different times in the year and over different periods of time, all of which affects the shape of the probability of flood function, O. As the decision rule for holding water in the reservoir depends on how the probability of flooding changes as the number of days to the irrigation season decreases and also depends on how the probability of flood changes as more water is stored, different flood probability functions will result in different quantities of water released with early snowmelt at different reservoirs. So, aside from the assumptions (stated above) that lead to the implication that the optimal threshold increases with time, t, we make no additional general assumptions about the shape of the function O.

2.1. *Simulation model*

Because the decisions to release and hold water under early snowmelt will be reservoir-specific, we create a simulation model of Ririe Reservoir, a small reservoir in southeastern Idaho that primarily provides flood control for the city of Idaho Falls. Ririe also stores irrigation water in a 'joint use' conservation storage and flood control pool for the greater Snake River project (recreation is not considered in our model, but Ririe is also used extensively for fishing and boating). The purpose of our simulation model is to determine the conditions, if any, under which early snowmelt leads to lowered irrigation deliveries. Ririe Reservoir impounds water from Willow Creek, a minor tributary of the Snake River. Willow Creek drains an area roughly 1500 km^2, 50% of which is at an elevation exceeding 1900 m above sea level. The reservoir has a total capacity of 99,300,000 m^3 available for flood control and irrigation, and the floodway system below the dam can carry flows up to 57 m^3/sec without causing flooding conditions downstream (US Bureau of Reclamation, 2012). Data for the period 1986 to 2011 from the US Geological Survey (USGS) suggest that snowmelt runoff exceeds this rate about once every 2.5 years. The same data also show that, while total capacity of Ririe Reservoir is just under 100 million m^3, average irrigation season deliveries are only 43 million m^3 – even in the heavier-than-average snow year of 2009, irrigation season deliveries only amounted to 55 million m^3 – putting Ririe in the category of reservoirs with irrigation obligations that represent only a fraction of total capacity.

We use streamflow data from USGS (from the stream gage on Willow Creek, just above Ririe Reservoir) to estimate the probability of flooding equation, O, for Ririe Reservoir. The probability of a flood is a function of two variables: the percentage of reservoir capacity filled and the number of days until the irrigation season begins (assuming that flood-level runoff always occurs before the start of irrigation). For a given percentage of capacity filled and a number of days until irrigation starts, the probability of a flood is the product[2] of the probability of the reservoir filling sometime before irrigation starts and the probability that a flood-quantity of water (i.e., more than 57 m^3/sec) enters the reservoir after that time. We estimate the probability of the reservoir filling a given amount of excess capacity in a given number of days by simulating 1600 10-day, 20-day, 30-day, 40-day and 50-day runoff scenarios, with initial capacity filled values between 85% and 99% of total available capacity. We use 10-day increments to simplify the calculation. The simulations are conducted by drawing N-day sequences randomly from the 25-year record of average daily streamflows during the runoff season (from March through May).

From these simulations, to determine the probability of flood in say, a 60-day period, with (100-ε)% of capacity remaining, we multiply the probability of the reservoir filling in 10 days (which is the number of observed simulations in which 10-day inflow exceeded [100-ε]% of capacity divided by 1600) by the probability that one flood would occur in the next 50 days, and add to that the product of the probability of filling in 20 days (but not 10 days, since the 10-day case is included in the 20-day case calculation) and the probability of a flood in the next 40 days and so on. Mathematically, we calculate

$$O(N, \varepsilon) = pr\left(\sum_{i=1}^{10} f_i \geq \frac{(100 - \varepsilon)}{100} S_{\text{max}}\right) \cdot pr(f_i \geq 57m^3/\text{sec}).(N - 10)$$

$$+ \left[pr\left(\sum_{i=1}^{20} f_i \geq \frac{(100 - \varepsilon)}{100} S_{\text{max}}\right) - pr\left(\sum_{i=1}^{10} f_i \geq \frac{(100 - \varepsilon)}{100} S_{\text{max}}\right)\right] \cdot pr(f_i \geq 57m^3/\text{sec}).(N - 20)$$

$$+ \ldots + \left[pr\left(\sum_{i=1}^{N-10} f_i \geq \frac{(100 - \varepsilon)}{100} S_{\text{max}}\right) - pr\left(\sum_{i=1}^{N-20} f_i \geq \frac{(100 - \varepsilon)}{100} S_{\text{max}}\right)\right] \cdot pr(f_i \geq 57m^3/\text{sec}).10$$

for $N = 60, 50, \ldots, 10$ and $\varepsilon = 83, 84, \ldots, 100$; and where $S_{max} = 99,300,000$ m^3, and f_i is flow rate of incoming water on a given day (the probability of inflow being greater than 57 m^3 is 0.0038) – as stated above, the f_i are simulated by drawing randomly from the daily average streamflow history for this stream gage. This gives 108 combinations of days and capacities with their associated simulated probabilities of flood, O. From those simulated probabilities, we estimate the following function (in log form) using Ordinary Least Squares (OLS): $O(\varepsilon, t) = b_1(T - t)^{b_2} \varepsilon^{b_3}$ we adjust notation slightly here, so that t is now the day of the year, and T is the day of the year when irrigation starts, so that $T-t$ is the number of days until the irrigation season begins.[3] Note that the assumptions we make in section 2 regarding the probability of flood function hold for any positive values for the parameters b_2 and b_3. Thus, the dam manager's decision rule to hold water, MB/D = $\partial O/\partial \varepsilon$, becomes

$$MB/D = b_3 b_1 (T - t)^{b_2} \varepsilon^{b_3 - 1} \tag{1}$$

The central question addressed in this paper is how early snowmelt affects the ability of reservoirs to provide irrigation water. To do so, we use the above decision rule to observe how a simulated dam manager's decision to hold water in Ririe Reservoir would differ in an average snowmelt year if the snowmelt were to arrive at the reservoir earlier, and calculate the difference in water available for irrigation between the two scenarios. We use the USGS stream gage information to determine a 'normal' runoff hydrograph (a graph of daily average runoff into the reservoir), as in Figure 1a. The 'normal' runoff is simply the average of all available annual hydrographs from this stream gage, and could be considered the expected runoff into the reservoir before any snowpack information or forecast is available. We then adjust the 'normal' hydrograph to various 'early snowmelt' scenarios (see Figure 1) in which the same quantity of runoff enters the reservoir, but with accelerated timing. Specifically, we create runoff hydrographs in which snowmelt runoff enters the reservoir four weeks and eight weeks early (simply shifting the runoff hydrograph 28 and 56 days to the left). We also consider the effects of snowmelt runoff coming earlier and in a shorter period of time. The average runoff hydrograph indicates that most of the runoff occurs in about a 100-day period. We create two new 'faster' runoff profiles in which the runoff occurs in a 70-day period: with peak runoff 28 days earlier than the average and with peak runoff 56 days earlier than average. We then run simulations with these five different runoff hydrographs (normal, 28 days early, 56 days early, 'faster' and 28 days early, 'faster' and 56 days early), in which the dam manager holds water during runoff according to the estimated decision rule, in Equation (1), and compare the quantity of water available for irrigation among the different scenarios.

To effectively use the decision rule (1), we need an estimate for the ratio of marginal irrigation benefits to the damage cost of a flood. We use a simplified estimate of damage cost, which is constant regardless of the size or timing of a flood. While a dam manager

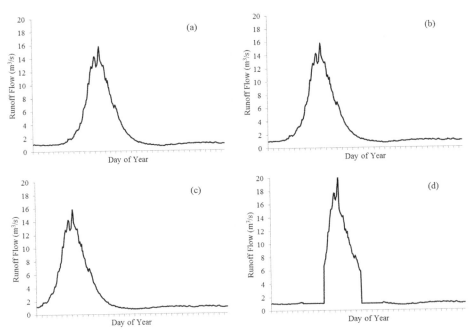

Figure 1. Hydrographs for Willow Creek entering Ririe Reservoir: (a) average daily inflow of the 35-year record, (b) inflow 28 days early, (c) inflow 56 days early, (d) inflow 28 days early and 'faster'.

would have more information about possible near-future streamflows in any given year (estimated from current snowpack information), which would adjust the expected cost of a potential flood up or down, we argue that the average used here is helpful for the purposes of simulation in which we're concerned with changes in an average year.

For the marginal irrigation benefit we use data from the 2007 USDA Census of Agriculture, which suggests that irrigated farms in Bonneville County, Idaho earn, on average, crop revenue[4] of $1837/ha. We assume, for simplicity, that the only option available to farmers who receive a lowered irrigation allotment is to irrigate less land, so the marginal benefit of additional irrigation water is the difference in average revenue of irrigated farms and the average revenue of dryland farms. In Bonneville county, the majority of dryland acreage is in wheat production, which earns crop revenue of $331/ha (Patterson & Smathers, 2004), making the marginal benefit of an additional acre of irrigated land equal to $1506. In addition to this value, we simulate the dam manager's decision with a marginal benefit value of $7079/ha, which is roughly equivalent to the difference in per-hectare revenue for high-yield potatoes and dryland wheat, and the use of which effectively assumes the highest possible marginal value of irrigation water. Use of this value is not entirely realistic, as one would assume that potatoes would be the last crop to be given up if irrigation water deliveries were diminished, but we include it for sensitivity analysis.

An anonymous reviewer noted that our assumption of no opportunity for producer adaptation and our use of average revenue for irrigated farms results in an over-estimate of the value of irrigation water. We therefore also run the simulation with a marginal benefit of additional irrigation water equal to the additional revenue from irrigated forage, which is the lowest-valued irrigated crop in Bonneville County. We choose this lowest-valued crop (in terms of producer revenue per acre) on the assumption that it would be the first crop abandoned for dryland farming if a farmer faced a reduced irrigation allotment. The associated marginal benefit is $563/ha.

The assumption of a constant damage cost of a flood, regardless of its size, restricts our model by limiting the expected damage cost of flood function (which is $D^*O(\varepsilon,t)$). To partially mitigate this shortcoming, we use varying values of the damage cost of a flood parameter, D. Specifically, we use values of $1,000,000 and $10,000,000. The higher value is about 40% of the cost (in 2012 dollars) of the 1962 floods of the Willow, Birch, Henry and Cedar Creeks – all of which are tributaries to the Snake River in Bonneville County – and inundated 56,000 acres during the flood event (Thomas & Lamke, 1962). This value can be considered a maximum damage cost estimate for a flood below Ririe Dam, considering that the flows that resulted in the 1962 floods are the highest in recorded history. We choose a lower bound on costs of $1,000,000, again, to test the sensitivity of the model to changes in the marginal benefit-cost ratio parameter. Simulating the model with the two different damage cost parameters should help to understand how this value affects the model outcomes, and how a true damage cost function, which varies with the size of a flood, might affect the dam manager's decisions and the consequences of early snowmelt.

3. Results

Our model simulation proceeds by, first, choosing an inflow scenario (normal, early, early and 'faster', etc.), reservoir capacity, initial capacity filled and marginal benefit-cost ratio. For each day of the simulated inflow, we calculate new quantity of water stored and the increase in probability of a flood as the reservoir fills (the right-hand side of Equation [1]). We compare the increase in probability of a flood to the marginal benefit-cost ratio, and determine the quantity of water to release. If the marginal benefit-cost ratio is greater or less than the increase in the probability of a flood, the inflow for that day is held in the reservoir or released, respectively, in accordance with the dam manager's decision rule. Once the irrigation season begins, the water is released as indicated by the average daily flows *out* from Ririe Reservoir, if there is enough water in the reservoir to provide those flows – that is, if capacity filled is greater than 1%. We then sum the quantity of daily average outflows from the reservoir to determine the quantity of water released during the irrigation season in a simulation run, which covers one runoff and associated irrigation season.

When we run the model with inflow with normal timing (i.e., no early snowmelt) the reservoir is able to provide 43,750,000 m³ of irrigation water (about 45% of the average annual inflow), which is roughly equal to the average irrigation-season deliveries from Ririe. Using the given capacity for Ririe Reservoir (99,300,000 m³), we also find that the modeled reservoir is able to provide the same quantity of irrigation water regardless of the inflow scenario, initial water storage or marginal benefit-cost ratio parameter chosen. As indicated in section 2 above, if the deliveries of irrigation water are significantly less than reservoir capacity, then the reservoir can meet those deliveries in a normal snowmelt year regardless of the timing of the snowmelt runoff – the reservoir either holds the runoff until irrigation starts (in a low initial storage scenario) or the reservoir releases the inflow during the runoff season, but is able to fulfill irrigation obligations with the water that was already in the reservoir (in a high initial storage scenario).

For the next set of simulations we adjust the total reservoir capacity downward to 61,650,000 m³ (about 60% of the original capacity), and leave the amount of snowmelt runoff that enters the reservoir unchanged, to see how irrigation deliveries are affected when runoff timing is shifted earlier. In Table 1, we report results for simulations with three different marginal benefit-cost ratios. The low marginal benefit-cost ratio corresponds to the lowest estimate of revenue of irrigated land and the highest estimate of damage cost of flood, while the high benefit cost ratio is calculated from the highest estimate of irrigated revenues and lowest flood damage cost. The intermediate value (in the center column) corresponds to high revenue, high damage cost scenario. In this case (where irrigation deliveries are 70% of reservoir capacity) early snowmelt results in less water being delivered for irrigation. Less water is supplied the earlier the snow melts and the lower the marginal benefit-cost ratio. When snow melts early, the quantity supplied ranges from 76% to 100% of normal deliveries, with the lowest quantity delivered in the case with the earliest, fastest snowmelt and the smallest benefit-cost ratio.

Table 1.   Simulation results – irrigation deliveries.

Runoff scenario	Irrigation deliveries		
	$MB/D = 0.0029$ (high benefits, low damages)	$MB/D = 0.000061$ (base benefits, high damages)	$MB/D = 0.000023$ (low benefits, high damages)
— m³ (% of Normal scenario) —			
Normal	43,746,840 (100%)	43,746,840 (100%)	43,746,840 (100%)
28 days early	43,746,840 (100%)	42,436,161 (97%)	40,779,009 (93%)
56 days early	39,463,398 (90%)	36,412,956 (83%)	35,018,433 (80%)
28 days early and 'fast'	43,621,074 (99.7%)	40,939,299 (94%)	39,176,109 (90%)
56 days early and 'fast'	37,439,829 (86%)	34,707,717 (79%)	33,035,769 (76%)

Table 2. Simulation results – dollar value of decreased irrigation deliveries.

Runoff scenario	Value		
	$MB/D = 0.0029$ (high benefits, low damages)	$MB/D = 0.000061$ (base benefits, high damages)	$MB/D = 0.000023$ (low benefits, high damages)
Normal	$0	$0	$0
28 days early	$0	$324,215	$274,398
56 days early	$4,976,505	$1,814,140	$807,006
28 days early and 'fast'	$146,115	$694,485	$422,598
56 days early and 'fast'	$7,327,488	$2,235,955	$990,318

The monetary cost of these lowered irrigation deliveries can be estimated by multiplying the difference in irrigation deliveries by some value per unit of water. The three different scenarios assume different values for irrigation water, however. Table 2 reports the value of lost expected revenue using the three different average revenue values ($563/ha, $1,506/ha, and $7,079/ha), the higher of which is likely to be unrealistic in estimating true costs of decreased availability of irrigation water.

4. Discussion

Early snowmelt does not appear to have a large effect on the quantity of water that is delivered from a reservoir with characteristics similar to the one modeled here. In some cases, the quantity of irrigation deliveries may be small enough, relative to the reservoir capacity (less than roughly 50%), that the reservoir will be able to either release the early-arriving snowmelt and have enough in storage to meet irrigation requirements, or the reservoir will have enough excess capacity to hold the early snowmelt without unduly increasing the probability of having to release a flood-quantity of water later. And even if the reservoir capacity is diminished relative to delivery obligations (something that could occur due to sedimentation or increased demand for water stored in the reservoir, for example), unless snowmelt arrives more than 28 days early, irrigation deliveries from the Ririe Reservoir are not likely to be significantly affected.

In only the most extreme cases, where snowmelt occurs eight weeks early (which is much earlier than the estimated 18 to 35 days caused by windborne dust), does early snowmelt cause irrigation deliveries to decrease by significantly. In these cases, it may be worthwhile to expend the effort required to determine the cause of the early snowmelt, and, if it is caused by windborne dust, how much of the dust that is accelerating the snowmelt is generated by local farmers. Models like the one presented in this paper may help to estimate costs of early snowmelt, but given that windborne dust can be carried long distances, a local solution (in which farmers who irrigate with water stored in the reservoir cooperate to control dust generation) may not be effective in restoring the runoff to its original timing. For example, a successful program to improve wind erosion of soil on farms in Bonneville County, or even all of Southeastern Idaho, may not improve hypothetical early snowmelt in the drainage above Ririe Reservoir if the majority of the dust deposited there originates in Northern Nevada or Utah.

A policy that is more likely to be effective is one that adjusts the dam manager's marginal benefit-cost ratio of holding water in the reservoir during spring runoff. This has the advantage of being effective despite the cause of early snowmelt (i.e., windborne dust and soot or global warming) and does not require coordination on the part of farmers who would have incentives to free-ride on others' efforts. The three parameters we identified that decreased irrigation deliveries were early snowmelt, 'faster' snowmelt and a lower benefit-cost value for holding water. Consider the irrigation deliveries between the simulations with different benefit-cost ratios: the results in Table 1 show that an order of magnitude increase of the marginal benefit-cost ratio can increase irrigation deliveries by 2% to 7% of the average delivery quantity. This suggests that, even without moving snowmelt timing closer to normal, irrigation water savings could be realized by decreasing the expected damage cost of flooding. Measures such as increasing width or depth of the floodway or outlet channel, or building infrastructure to create possible bypass diversions to other watersheds would increase the quantity of flood water that is considered a 'flood' in the dam manager's calculation of flooding probability and/or would decrease the cost of a potential flood that exceeded that quantity. Either of these effects would result in the dam manager holding a greater maximum quantity of water earlier in the season.

Decreasing development within the historical floodplain below the dam to decrease the damage cost of flooding is another potentially effective policy, although one with a wholly different set of costs and additional ecological benefits. Development for residential and commercial properties is much more lucrative than farming, and would be difficult to advocate foregoing for a small percentage increase in farm revenues. This is especially true given the results of the simulation, which indicate that early snowmelt causes the large reductions in irrigation deliveries only when the marginal benefits of water for irrigation, and the resulting costs of early snowmelt, are assumed to be small.

Notes

1. We note that this optimization framework assumes risk-neutrality for the dam manager. The analysis in this paper therefore may underestimate the actions taken by a risk-averse decision maker to minimize the expected cost of flooding.
2. Those two events (the reservoir filling and a flood occurring afterwards) are not likely to be independent, but the conditional probabilities between the two are not known and are difficult to estimate, and would complicate the calculation of the function O. With the calculation described, we effectively over-estimate the lowest probabilities of flooding and under-estimate the highest probabilities, somewhat decreasing the variance in O.
3. This functional form fits the simulated probability of flood fairly well for these historical data for this particular drainage – it may or may not be applicable in general. The estimated equation is $\exp(-105.732[13.00])(T - t)^{4.82[0.269]}\varepsilon^{18.84[2.87]}$, with standard errors of the parameter estimates in brackets. The R^2 of the regression is 0.77.
4. We use total revenue, rather than net revenue to measure benefits of irrigation water. If the dam manager wishes to maximize irrigator benefits, then this formulation overstates the benefits of irrigation since the costs of dryland farming are likely lower than costs of irrigated farming.

References

Chatterjee, B., Howitt, R. E., & Sexton, R. J. (1998). The optimal joint provision of water for irrigation and hydropower. *Journal of Environmental Economics and Management, 36*, 295–313.

Dudley, N. (1988). A single decision-maker approach to irrigation, reservoir and farm management decision making. *Water Resources Research, 24*, 633–640.

Heidecke, C. & Heckelei, T. (2010). Impacts of changing water inflow distributions on irrigation and farm income along the Drâa River in Morocco. *Agricultural Economics, 41*, 135–149.

Lee, Y., Yoon, T., & Shah, F. A. (2011). Economics of integrated watershed management in the presence of a dam. *Water Resources Research, 47*, W10509. doi:10.1029/2010WR009172.

Leung, L. R., Qian, Y., Bian, X., Washington, W. M., Han, J., & Roads, J. O. (2004). Mid-century ensemble regional climate change scenarios for the western United States. *Climatic Change, 62*, 75–113.

Neff, J. C., Ballantyne, A. P., Farmer, G. L., Mahowald, N. M., Conroy, J. L., Landry, C. C., . . . Reynolds, R. L. (2008). Increasing eolian dust deposition in the western United States linked to human activity. *Nature Geosciences, 1*, 189–195. doi:10.1038/ngeo133

Painter, T. H., Barrett, A. P., Landry, C. C., Neff, J. C., Cassidy, M. P., Lawrence, C. R., . . . Farmer, G. L. (2007). Impact of disturbed desert soils on duration of mountain snow cover. *Geophysical Research Letters, 34*, L12502. doi:10.1029/2007GL030284

Palmieri, A., Shah, F., & Dinar, A. (2001). Economics of reservoir sedimentation and sustainable management of dams. *Journal of Environmental Management, 61*, 149–163.

Patterson, P. E. & Smathers, R. L. (2004). Winter wheat production costs and budgeting. In L. D. Robertson, S. O. Guy, & B. D. Brown (Eds.), *Southern Idaho dryland winter wheat production guide* (pp. 67–74). Moscow: University of Idaho Bulletin 827.

Qian, Y., Gustafson, W. I. Jr., Leung, L. R., & Ghan, S. J. (2009). Effects of soot-induced snow albedo change on snowpack and hydrological cycle in western United States based on Weather Research and Forecasting chemistry and regional climate simulations. *Journal of Geophysical Research, 114*, D03108. doi:10.1029/2008JD011039

Rowell, D. P. (2005). A scenario of European climate change for the late twenty-first century: Seasonal means and interannual variability. *Climate Dynamics, 25*, 837–849.

Seneviratne, S. I., Lüthi, D., Litschi, M., & Schär, C. (2006). Land-atmosphere coupling and climate change in Europe. *Nature, 443*, 205–209.

Steltzer, H., Landry, C., Painter, T. H., Anderson, J., & Ayres, E. (2009). Biological consequences of earlier snowmelt from desert dust deposition in alpine landscapes. *Proceedings of the National Academy of the Sciences, 106*, 11629–11634.

Stewart, I. T., Cayan, D. R., & Dettinger, M. D. (2004). Changes in snowmelt runoff timing in Western North America under a 'business as usual' climate change scenario. *Climatic Change, 62*, 217–232.

Thomas, C. A. & Lamke, R. D. (1962). *Floods of February 1962 in Southern Idaho and Northeastern Nevada. Geological Survey Circular 467*. Washington, DC: United States Department of the Interior.

United States Bureau of Reclamation. (2012). *Project details – Ririe Project* [online]. Washington, DC: United States Department of the Interior. Retrieved from http://www.usbr.gov/projects/Project.jsp?proj_Name=Ririe%20Project&pageType=ProjectPage

United States Department of Agriculture. (2007). "Census of Agriculture." Washington, DC: National Agricultural Statistical Service.

Vedula, S. & Kumar, D. N. (1996). An integrated model for optimal reservoir operation for irrigation of multiple crops. *Water Resources Research, 32*, 1101–1108.

Vedula, S. & Mujumdar, P. P. (1992). Optimal reservoir operation for irrigation of multiple crops. *Water Resources Research, 28*, 1–9.

Climate change opportunities for Idaho's irrigation supply and deliveries

Russell J. Qualls[a], R. Garth Taylor[b], Joel Hamilton[b] and Ayodeji B. Arogundade[a]

[a]*Department of Biological & Agricultural Engineering, P.O. Box 440904, University of Idaho,*
[b]*Agricultural Economics & Rural Sociology Department, University of Idaho*

The Snowmelt Runoff Model (SRM) was used to simulate timing and magnitude of runoff for six climate scenarios (2030 and 2080 'Wet', 'Middle', and 'Dry'). The water supply results from SRM were run through a Southern Idaho reservoir operation and water rights allocation model (MODSIM). The 2030-Dry and 2080-Dry scenarios produce supply deficits relative to the current climate of 5.4%, and 1.9%, respectively, for which the corresponding irrigation water delivery reductions were 1.7% and 2.7%. In contrast, the 2030-Wet, 2030-Mid, 2080-Wet, and 2080-Mid climate change scenarios increased water supply by 13.4%, 0.5%, 19.5%, and 5%, respectively, for which water deliveries increased by 0.41%, 0.04%, 0.34% and 0.14%, respectively. Idaho's irrigation delivery and storage system can ameliorate the risk of dry climate change, but is incapable of storing and delivering the increased water supplied by the wet climate change scenarios. This is an opportunity worth exploring.

1. Introduction

Idaho's $7.7 billion agricultural industry is the third largest in the west, behind California and Washington (Eborn, Patterson, & Taylor, 2012). Idaho's agricultural productivity is centered in the desert of the Snake River Plain and survives on irrigation water from the Snake River. Idaho is second to California in total irrigation water withdrawals (Kenny et al., 2009). Of the total 18.6 million acre-feet (AF) (1 AF = 1,233 m^3) of Idaho irrigation withdrawals, 14.2 million AF and 4.3 million AF are withdrawn from surface and ground-water sources, respectively. Snake River headwaters are fed by snow pack in the mountains of western Wyoming, southwestern Montana, and far eastern Idaho. The runoff is stored in large reservoirs for irrigation of southern and eastern Idaho.

Hydrologic processes of a region determine only a portion of the net effect of climate variability. Highly managed systems, such as the Snake River Basin, store, release, and divert water to modify the spatial and temporal movement of water within and between regions. In addition to unique climate, hydrology, and physical infrastructure, the system is governed by an institutional framework unique to each state. In Idaho, as with all western states, the institutional foundation is water rights governed by the Doctrine of Prior Appropriations.

Much of the climate change research in snowmelt dominated regions has focused primarily or exclusively on the hydrologic changes without consideration of management or legal influences (e.g., Gillan, Harper, & Moore, 2010; Mote, Hamlet, & Salathe, 2008; Stewart, Cayan, & Dettinger, 2004). In this research, climate change is examined from the perspective of long run impacts on the supply and delivery of irrigation water to southern Idaho agriculture. Specifically, climate change induced modifications to irrigation water supply were examined from a legal perspective that measures whether climate induced water supply changes satisfy, exceed, or fall short of the appropriated allocation. The legal approach can be used to examine adjudication issues that may arise as climate change alters runoff and is a critical first step in water planning to address remediation alternatives such as construction of new or expanded storage.

2. Methods and data

To project the impacts of climate change on the water supply and distribution across the Snake River Plain of southern Idaho, the methods proceeded in two steps. Annual snowmelt runoff hydrographs were produced by the Snowmelt Runoff Model (SRM). These supply scenarios were then input into MODSIM (Labadie, 1995) which simulates the physical and legal distribution of water across the Snake River basin.

The study region is the Snake River basin from its headwaters in Yellowstone and Teton National Parks in Wyoming to Brownlee Reservoir in Hells Canyon on the border between Idaho and Oregon. This region encompasses over three million irrigated acres scattered across southern Idaho, about one half of the irrigated acreage in the Pacific Northwest. Over half of the southern Idaho water supply is snowmelt-runoff from the mountains in the eastern part of the study region. The South Fork of the Snake River basin drains 3465 mi^2 (8974 km^2) and ranges in elevation from 5799 to 13760 feet (1768 to 4194 m) above mean sea level (Seaber, Kapinos, & Knapp, 1987) and supplies more than one-third of the total annual volume of flow entering Brownlee Reservoir (Figure 1). The Snake River basin is divided into three reaches to which the Boise Payette river system is added near Brownlee Reservoir. The uppermost reach, above American Falls is allocated over 40% of the irrigation water (Table 3). The middle reach from American Falls to King Hill is allocated close to a third of the water. Below King Hill has little irrigated agriculture dependent on Snake River water. An endangered species fish flow allocation of 427 thousand acre feet (KAF) is about 3% of the basin flow.

2.1 Snowmelt runoff model

Climate integrates long term cycles associated with solar output, greenhouse gas, and atmospheric reflection of radiation, together with periodicity associated with El Nino/La Nina cycles or Pacific Decadal Oscillation and annual drought and flood events. To model the range of plausible climate scenarios for the Snake River basin, Atmosphere-Ocean General Circulation Model (AO-GCMs) simulations were superimposed onto the historically observed climate variability.

SRM (Martinec, Rango, & Roberts, 2005) is a daily time step degree-day model. Runoff is aggregated from snowmelt generated from the fraction of area within a basin covered by snow on a given day. SRM data inputs are time series of precipitation, air temperature, and basin snow cover. Model output is year-round daily time series of stream flow from snowmelt runoff.

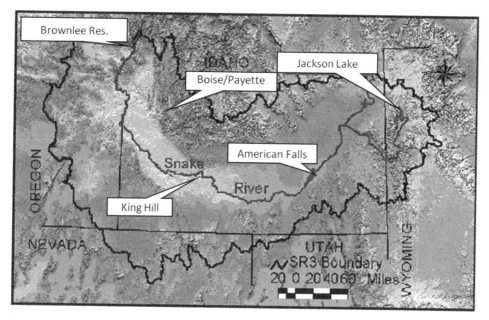

Figure 1. (Color online) Map of the Snake River Drainage from Jackson Lake in Wyoming to Brownlee Reservoir in southwestern Idaho, denoted by the 'SR3 Boundary' line, bordering Montana, Wyoming, Utah, Nevada, and Oregon. MODSIM reach demarcations are shown.

SRM includes a 'climate change' component which allows modification of the input data according to prescribed climate change characteristics, such as temperature or precipitation changes for specified period(s) during the year. In the climate change mode, SRM modifies snow-covered area to accommodate warming and resulting snowmelt changes. SRM has been applied to a range of watershed sizes (e.g., Dey, Sharma, & Rango, 1989; Gomez-Landesa & Rango, 2002; Hong & Cheng, 2003; Zhang et al., 2007).

The daily average air temperature and precipitation data for SRM for the Snake River headwaters (Qualls & Arogundade, 2012) were collected from 11 automated Snow Telemetry stations (SNOTEL) within the South Fork headwaters (USDA Natural Resources Conservation Service [NRCS]). Localized data were required to downscale the climate scenario temperature and precipitation perturbations to the study area. Images from the Moderate Resolution Imaging Spectroradiometer (MODIS) instrument aboard the NASA Earth Observing System satellite 'Terra' (Hall, Riggs. & Salomonson, 2006) were used to derive snow-covered-area curves. The Snake River SRM was validated using historical daily average streamflow data taken from the Upper Snake River Basin (United States Geological Survey [USGS] National Water Information System) and corrected for upstream storage (US Bureau of Reclamation [USBR]).

Given the uncertainty of future climate on water supply, a range of climate change scenarios were generated. Scenarios were selected from 18 different models run for the Intergovernmental Panel on Climate Change Fourth Assessment Report (IPCC, 2007). Each model was run using a variety of carbon emissions scenarios developed by the IPCC. GCM results corresponding to the A1B emissions scenario were used. The A1B scenario represents a declining emphasis on carbon emitting energy sources until 2050, and balanced emphasis on carbon and non-carbon emitting sources thereafter.

Table 1. Idaho GCMa precipitation and temperature climate scenarios.

	Wet		Middle		Dry	
	Canadian (cccma.t63)		NCAR (PCM)		NOAA (gfdl0)	
	Precipitation (%)	Temperature (°C)	Precipitation (%)	Temperature (°C)	Precipitation (%)	Temperature (°C)
2030 (2020–2039)						
January	−15.99	1.21	−2.33	0.99	−6.51	0.32
February	23.16	1.93	4.43	0.77	−10.87	1.37
March	13.28	1.21	0.75	0.73	1.19	1.95
April	11.63	0.85	−0.12	0.47	11.80	1.03
May	21.03	1.44	7.41	0.79	1.62	0.21
June	18.01	0.61	7.99	0.83	−17.70	0.15
July	−2.92	0.88	−4.43	1.11	−32.36	3.17
August	15.95	0.44	3.92	0.90	−47.88	3.51
September	9.26	0.01	−17.01	1.28	0.38	1.68
October	17.15	1.10	13.87	0.80	−3.71	1.95
November	22.37	0.49	−3.92	0.55	0.81	1.51
December	22.83	1.71	−4.19	1.03	−1.82	1.20
Annual	12.93	0.99	1.16	0.86	−7.64	1.50
2080 (2070–2089)						
January	4.52	3.38	6.95	3.84	−10.69	3.34
February	37.30	4.40	14.12	3.60	6.60	4.61
March	24.66	2.58	9.35	1.64	14.44	3.98
April	36.00	2.58	10.85	1.51	19.35	2.98
May	20.94	3.33	11.00	1.60	2.38	2.10
June	−6.83	2.85	11.79	2.50	−26.66	2.83
July	−9.16	3.08	−4.90	2.97	−50.35	7.50
August	−7.66	2.44	8.13	2.82	−47.16	8.52
September	16.70	2.50	−29.77	3.30	−40.71	5.48
October	13.25	3.04	12.96	2.26	−18.14	4.28
November	36.41	2.10	1.46	1.99	6.71	3.68
December	23.48	3.93	−3.90	2.72	31.55	3.25
Annual	17.41	3.02	5.32	2.56	−6.75	4.38

Note: [a]GCM models used were Canadian Centre for Climate Modeling & Analysis (cccma.t63), National Center for Atmospheric Research (PCM), and US Department of Commerce/ NOAA/Geophysical Fluid Dynamics Laboratory (gfdl0) for the wet, middle and dry scenarios respectively.

GCM scenarios show greater variability in precipitation than in temperature. Three specific GCMs were selected to capture a wide range of precipitation change, similar to the 10%, 50% and 90% probability density function quintiles from the output of the 18 GCMs run for the IPCC (see Table 1). Precipitation and temperature scenario data covering a 500 km area of Southern Idaho and Southwestern Montana were extracted from GCM model output. The climate scenarios were downscaled using SNOTEL data and distributed satellite data because the GCM output did not account for local differences resulting from topography.

The climate change scenarios simulate the periods 2020 to 2039 (2030) and 2070 to 2089 (2080) relative to a climate base period of 1980 to 1999. Each change scenario included a percentage change to monthly precipitation, and degree change to monthly temperature, for the 12 months of the year, relative to the base period. Thus, six scenarios were simulated, a dry, average, and wet precipitation scenario for the 2030 and

2080 time periods (Table 1). The GCM output provided monthly averages, not changes in daily or inter-annual variability. The GCM output was combined with observed SNOTEL weather data surrounding the climate base period by adding the temperature changes to the observed record (e.g., adding the monthly January temperature increase from a particular scenario to each day of the observed January data), and scaling a given month's observed daily precipitation record by the corresponding month's GCM percentage change in precipitation. Thus, the historic daily and inter-annual weather variability from the 24-year observed SNOTEL data record were preserved, while incorporating the climate scenario perturbations to temperature and precipitation on a monthly time step. This generated a 24-year perturbed daily climate for each climate scenario for snowmelt runoff modeling and production of hydrographs.

2.2 Water allocation model MODSIM

Beginning with Maass et al. (1962), hydrologic systems have been modeled as networks of water storage, demand, and supply nodes linked by the conveyance structures of rivers, canals, and pipelines. Linked hydrologic economic models take either of two design options, simulation or optimization (Harou et al., 2009). This research takes a network simulation approach that conjoins a hydrologic network model, that is, a river-reservoir prior appropriations water allocation model (MODSIM, Labadie, 1995) with a system-wide agricultural optimization model (Hamilton, Green, & Holland, 1999; Houck et al., 2007).

The USBR's Snake River Basin MODSIM model was used to examine the effects of climate change and variability on water allocations across the Snake River basin – from headwaters above Jackson Lake west to the Oregon/Idaho border (Figure 1). MODSIM is a generalized river basin decision support system and network flow model (Labadie, 1995). Using a monthly time step, MODSIM routes water through basin streams and storage reservoirs, using appropriation doctrine priorities, to allocate available water to satisfy irrigation demands and minimize shortages. MODSIM distributes the inflows throughout the system among nodes and users by means of the links, satisfying limitations imposed by water rights, timing of inflows, and storage and transport capacities. MODSIM has 96 nodes representing allocations within the Snake River Plain. Nodes are individuals or collections of users such as irrigation districts. Each node contains information on the water inflows, reservoir operating rules, and irrigation demands calibrated for a 62-year baseline of historic inflows. The links contain information on water right priorities and delivery system capacities. The Snake River Basin MODSIM tracks only irrigation water usage, Idaho's minuscule municipal and industrial water use are ignored. The irrigation exception is the inclusion of the 427 KAF endangered species fish flow augmentation at the Oregon border.

When forecasting future water supply and demand conditions, a common problem with basin water allocation models is infeasible solutions (Michelsen & Taylor, 1999). MODSIM does not have constraints that produce infeasible solutions. Rather, available water supplies are allocated by seniority of appropriated water right. MODSIM allocates available water supplies to meet seniority of appropriated water right while respecting water right priorities and system capacities. Supply shortfalls, such as those resulting from climate change, result in the lower priority water demand going unfulfilled. MODSIM evaluated the potential effects of climate change on water allocation, with the inflow data at each of the system's spatially distributed input nodes modified by the monthly percentage changes to inflow simulated by SRM, for each of the six climate change scenarios.

The output provides the annual historic water usage at each node, the volumetric delivery for the 1980 to 1999 climate base period and for each of the climate change scenarios, as well as the volumetric shortage for the base climate and for each of the climate change scenarios. MODSIM specifies the amount of water delivered to each demand node.

3. Results and discussion

3.1 Snowmelt runoff modeling

The climate scenario simulated hydrographs from SRM incorporate both long-term climate change as well as inter- and intra-year variations. Monthly perturbations within each climate scenario are static from year to year, whereas inter-annual variability comes from the historical dataset (i.e., differences among years within the 24-year SRM historical dataset). The historical inter-annual variability is large (i.e., the largest annual flow volume is three times the smallest) and changes caused by the climate scenarios are much smaller (-5.4% to $+19.5\%$). The historical inter-annual variability is filtered out to emphasize the long-term impact of the climate scenarios by averaging the SRM hydrograph for each day of the water year over the 24-year historical data set.

The average annual snowmelt runoff hydrographs for the South Fork of the Snake River are shown in Figure 2 (2030 simulations) and Figure 3 (2080 simulations). The accompanying annual average percentage changes to total water supply corresponding to each climate scenario appear in Table 2. Average daily flow measurements and simulations of the historical climate base period shown in Figures 2 and 3, are referenced as 'Measured' and 'Simulated', respectively. Simulated flows for the six climate change scenarios are referenced by the particular climate change scenario (e.g., '2030-Wet').

The climate scenario simulations change the South Fork of the Snake River hydrographs in two aspects: (1) Total annual volume of discharge; and (2) within-year timing

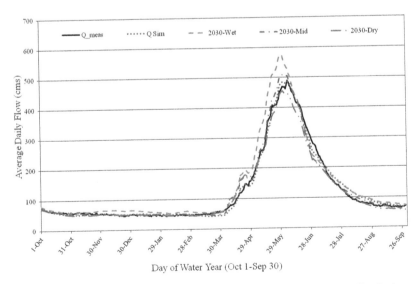

Figure 2. (Color online) Comparison of the 2030 climate change simulated daily discharges with 'natural' streamflow observations for the climate base period, corrected for upstream reservoir storage (Q_meas) and climate base period simulated (Q sim) discharges averaged over a 24-year period.

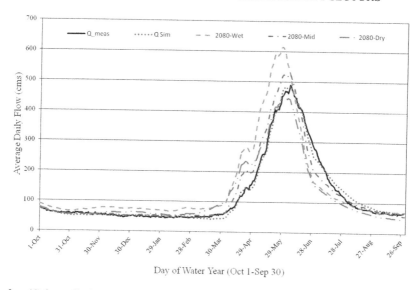

Day of Water Year (Oct 1-Sep 30)

Figure 3. (Color online) Comparison of the 2080 climate change simulated daily discharges with climate base period, measured and simulated discharges averaged over a 24-year period. (See Figure 2 for definitions).

Table 2. Comparison of changes associated with six climate scenarios relative to simulated historical flows.

Scenario	SRM Water Supply[a] % Change	MODSIM Delivery[a] % Change	Peak Advance[b] (Days)	V_50 Advance[c] (Days)
2030-Wet	13.4	0.41	6	5
2030-Mid	0.5	0.04	5	3
2030-Dry	−5.4	1.69	5	3
2080-Wet	19.5	0.34	8	14
2080-Mid	5	0.14	7	6
2080-Dry	−1.9	−2.73	7	13

Note: [a]SRM Water Supply and MODSIM deliveries are volumetric percentage changes.
[b]Peak advance is the number of days that the peak of the hydrograph moves earlier in the melt season relative to the historical simulation.
[c]V_50 is the number of days earlier that 50% of the total annual volume flows out of the basin relative to the historical simulation.

of runoff. The Wet and Middle scenarios show a net annual increase in the volume of flow or water supply for both the 2030 and 2080 simulation periods. These increases from Table 2 are +19.5% (2080-Wet), +13.4% (2030-Wet), +5% (2080-Mid), and 0.5% (2030-Mid). Both the 2030 and 2080 simulation periods for the Dry model show a net decrease in annual flow volume, with a slight decrease of 1.9% for the later 2080-Dry scenario, and a larger 5.4% decrease for the earlier 2030-Dry scenario (Table 2). These increases or decreases to the annual discharges reflect the percentage changes to precipitation of the corresponding climate change scenarios.

In addition to changes in the annual volume of flows, there is a change in the timing of snow melt and stream flow (Figures 2 and 3). The timing of the simulated historical values closely corresponds to the timing of the measured flows. Consistent with other Pacific Northwest research (e.g., Stewart et al., 2004), all the climate change scenarios advance the spring runoff in time. The hydrograph peaks advance by five to eight days relative to the simulated historical flows, with 2080 climate scenario peaks occurring earlier than those of the 2030 scenarios (Table 2). The date by which 50% of the total annual volume of runoff has occurred (V_{50}, Table 2) advances between three and five days earlier for each of the 2030 scenarios, six days for the 2080-Mid scenario, and 13 to 14 days for the 2080-Wet and 2080-Dry scenarios relative to the historical case. Corresponding to a shift toward earlier spring runoff, there is a reduction in summer runoff in several of the changed-climate scenarios, especially those for 2080. Timing shifts are the result of increased temperatures in every month of every climate scenario (see Table 1) and manifest themselves as increased winter-time rainfall as evidenced by the increased winter flows, especially in the two 'wet' scenarios, and the advance of spring snowmelt. The 2080-Wet and 2080-Dry scenarios have springtime (April, May, June) temperatures 2 to 3°C warmer than the base climate. This warming increases the spring runoff, and melts the snowpack earlier, reducing summer flows.

3.2 MODSIM results

Water supply changes from the SRM simulations of the climate scenarios were fed into the managed Snake River modeling system MODSIM to determine the magnitude of deliveries. The two dry scenarios produce deficits in water delivery relative to the base period climate, of 1.69% for the 2030-Dry scenario, and 2.73% for the 2080-Dry scenario (Table 3). In contrast, the other four climate change scenarios (2030-Wet, 2030-Mid, 2080-Wet, and 2080-Mid) increase water delivery relative to the base period climate by 0.41%, 0.04%, 0.34% and 0.14%, respectively.

In SRM, the total annual supply was reduced by 5.4% and 1.9% in the 2030-Dry and 2080-Dry scenarios, respectively (Table 3). In contrast, with MODSIM the reduction in deliveries was less for the 2030-Dry climate scenario than for 2080-Dry (-1.69% versus -2.73%, Table 3). The interaction of spring runoff timing with the storage infrastructure reverses the SRM and MODSIM relative severities of the 2030- and 2080-Dry scenarios. Although the 2030-Dry scenario reduces annual water supply by more than the 2080-Dry scenario, the April-May-June temperature changes associated with the 2030-Dry scenario, ranging from +0.15 to +1.03°C (Table 1), are not warm enough to increase rate of spring snowmelt so that runoff exceeds system reservoir capacity. Further, preserving spring snowpack maintains higher summer flows. As a result, a greater proportion of runoff can be stored to meet the irrigation season allocation. In contrast, in the 2080-Dry scenario the early spring temperature changes, ranging from +2.10 to +3.98°C between March and May (Table 1), increase the early spring snowmelt and cause some low elevation precipitation to fall as rain rather than snow so that runoff exceeds reservoir storage capacity. Water that could otherwise be stored is spilled and the rapidly disappearing snowpack reduces summer flows. Inability to store during the spring, and lower summer flows exacerbate the supply shortage for the 2080-Dry scenario.

The contrast between the SRM supply and MODSIM allocation results for the 2030-Wet and 2080-Wet climate scenarios is even more pronounced. In these two scenarios respectively, SRM shows a mean annual water supply increase of 13.4% and 19.5%, but MODSIM shows deliveries increase by only 0.41% and 0.34% (Table 2). Despite large increases in snowmelt runoff due to increased precipitation, the system does not have the capacity to allocate or store the increased runoff.

The spatial distribution of MODSIM results across the Snake River Plain are summarized in Table 3, grouped into three sections: A. Demand and Delivery; B. Demand and

Table 3. MODSIM water-right allocations (demands), delivery, shortages and delivery changes.

A. Demand and Delivery (million AF)

River Segment	Demand (million AF)	Delivery (million AF)						
		Base	2030-Wet	2030-Mid	2030-Dry	2080-Wet	2080-Mid	2080-Dry
Snake River Above American Falls	5.097	5.061	5.077	5.057	4.958	5.075	5.060	4.877
American Falls to King Hill	3.965	3.954	3.964	3.956	3.926	3.963	3.960	3.918
Snake River Below King Hill	0.522	0.507	0.516	0.510	0.500	0.517	0.511	0.498
Boise and Payette Rivers	2.620	2.504	2.611	2.602	2.558	2.612	2.602	2.528
In-Stream Fish Flow Water	0.427	0.416	0.426	0.422	0.387	0.418	0.424	0.378
Basin Total	12.631	12.542	12.593	12.547	12.329	12.585	12.558	12.198

B. Demand and Shortage

River Segment	Demand (million AF)	Shortage (thousand AF)						
		Base	2030-Wet	2030-Mid	2030-Dry	2080-Wet	2080-Mid	2080-Dry
Snake River Above American Falls	5.097	37	21	41	140	23	38	221
American Falls to King Hill	3.965	11	1	8	38	2	5	47
Snake River Below King Hill	0.522	15	6	12	22	5	10	24
Boise and Payette Rivers	2.620	16	8	17	61	8	17	91
In-Stream Fish Flow Water	0.427	11	1	6	40	9	3	49
Basin Total:	12.6	89	38	84	302	46	72	433
Shortage as % of Demand		0.7	0.30	0.66	2.39	0.37	0.57	3.42

C. Delivery Relative to Base Delivery (% change)

River Segment	# Nodes	Base	2030-Wet	2030-Mid	2030-Dry	2080-Wet	2080-Mid	2080-Dry
Snake River Above American Falls	49	0	0.31	-0.08	-2.04	0.28	-0.02	-3.64
American Falls to King Hill	19	0	0.26	0.07	-0.69	0.23	0.16	-0.91
Snake River Below King Hill	2	0	1.76	0.53	-1.47	1.88	0.85	-1.84
Boise and Payette Rivers	25	0	0.28	-0.06	-1.76	0.31	-0.05	-2.91
In-Stream Fish Flow Water	1	0	2.42	1.43	-6.85	0.59	2.08	-9.08
Basin Total	96	0	0.41	0.04	-1.69	0.34	0.14	-2.73

Note: 1 AF = 1,233 m^3

Shortage; and C. Climate Change Scenario Delivery Relative to Base Climate Delivery. Results in each section are aggregated into four major reaches of the Snake River (from east to west: Above American Falls, American Falls to King Hill, Below King Hill, and Boise and Payette Rivers), and In-Stream Fish Flow.

Table 3, Section A shows the agricultural water deliveries in the basin, for full provision of historical allocated water rights (hereafter 'demand') and for the base climate and climate change scenarios. Demand represents the annual average allocations associated with the 62-year historical inflow database of MODSIM. The base climate deliveries represent the average annual actual deliveries for the 1980 to 1999 climate that satisfy the prior appropriation water rights. The remaining columns represent the average annual deliveries for the respective climate change scenarios. The base climate and all climate change scenarios generate annual average water deliveries that are lower than the full demand. While the full allocation is met in some individual years, the allocation is not satisfied by the average flows of the base climate or any of the climate change scenarios.

The annual average shortage for a given scenario is the difference between delivery and demand (Table 3, section B). Total shortages combine the effects of over-allocation of water rights relative to the base climate and the shortage resulting from climate change. With the base climate, the shortage is 0.7% of allocated demand, that is, the basin is over-allocated by less than 1%. Over 40% (37/89, Table 3) of the base climate shortage occurs above American Falls where over 40% of the basin water is allocated. The shortages for the 'wet' and 'middle' cases for both the 2030 and 2080 time periods are miniscule, ranging from 0.3% of demand for the 2030-Wet scenario to 0.66% of demand for the 2030-Mid scenario (Table 3). The shortages for the 2030-Dry and 2080-Dry scenarios are 2.4% and 3.4% of demand, respectively.

In Section C of Table 3, the effect of over-allocation is removed so that only the effects of the climate change scenarios relative to the base climate are considered. The basin total amounts from Section C correspond to MODSIM delivery percentage changes from Table 2. Positive numbers indicate the climate change scenario provides more water, or satisfies the allocation more fully than the base climate, and conversely for negative numbers. The four Wet- and Mid- scenarios result in slight excess deliveries relative to the base climate delivery and the Dry- climate scenarios produce minor deficits relative to base climate delivery.

MODSIM deliveries can be grouped by nodes according to ranges of percentage change in delivery relative to base climate deliveries (Table 4). For the climate scenarios with large supply increases ranging from 5% to 19.4% (2030-Wet, 2080-Wet and 2080-Mid), none of the 96 nodes receives a delivery increase to match the supply increase. For the 2030-Dry and 2080-Dry scenarios, there are 13 and 37 nodes, respectively, which have a greater reduction in delivery than the corresponding supply reduction (-5.4% and -1.9%; Table 4).

Shortages or surpluses are not uniformly distributed across nodes as illustrated by the distribution of percentage change to base climate delivery against cumulative base climate delivery (Figure 4). For each climate scenario, Figure 4 shows how many AF are subject to a given percentage shortage relative to base climate deliveries. The node-level data are sorted in descending order of percentage shortage, with the largest shortages at the left and base climate deliveries incrementally added along the horizontal axis, so that the nodes experiencing surpluses appear at the far right. For 2030-Dry and 2080-Dry, nodes accounting for a total of about 1.4 million acre feet (MAF) (11%) and 2.0 MAF (15%) of base climate deliveries, respectively, are subject to shortages greater than 5% (Figure 4). Most nodes, however, experience shortages less than 5% (Figure 4). The other scenarios (wet

Table 4. Number of MODSIM nodes within a range of percentage change relative to base deliveries.

% Change in Deliveries[a]	2030-Wet	2030-Mid	2030-Dry	2080-Wet	2080-Mid	2080-Dry
<-15	0	0	1	0	0	5
(-10,-15]	0	0	1	0	0	5
(-5,-10]	0	0	16	0	0	11
(0,-5]	5	34	51	12	27	41
[0]	45	43	26	43	43	31
(0,5)	43	19	1	38	26	3
[5,10)	2	0	0	2	0	0
[10,15)	1	0	0	1	0	0
>=15	0	0	0	0	0	0
Total Nodes	96	96	96	96	96	96
% Change in Supply (SRM)	13.4	0.5	−5.4	19.5	5.0	−1.9
Nodes with Delivery Chg < Supply Chg	96	90	13	96	96	37

Note: [a]Parentheses '(' indicate non-inclusive end of range.
Brackets '[' indicate inclusive end of range.

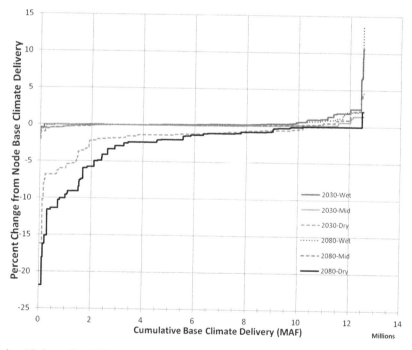

Figure 4. (Color online) Climate scenario impact on climate base period deliveries at nodes within MODSIM expressed as a percentage of base climate delivery and aggregated across the cumulative delivery. Lengths of horizontal line segments represent incremental volume of base climate delivery at nodes (e.g., the node at the left end of 2080-Dry line experiences a 22% reduction from base climate delivery of about 96 KAF; 1 AF=1,233 m^3).

and mid) all have nodes which experience small shortages even though SRM modeled a net surplus of water for those scenarios. The shortages and surpluses for the wet and mid scenarios are similar enough that they are difficult to distinguish.

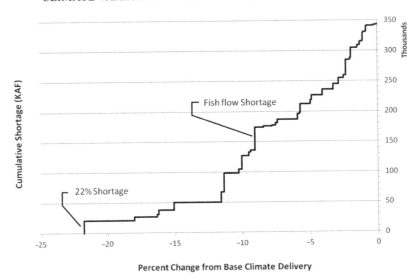

Figure 5.   2080-Dry MODSIM delivery shortages (AF; 1 AF=1,233 m³) expressed in terms of percentage change from base climate delivery. Total 2080-Dry shortage is 343 KAF. Vertical heights of line segments indicate magnitude of shortage at that percentage change. Examples: (a) 20 KAF shortage occurs at node with 22% reduction; (b) the fish flow shortage segment has a 9% shortage or 38 KAF (174–136 KAF).

In the two dry climate scenarios, shortages are concentrated on a relatively small number of severely impacted nodes. To illustrate, Figure 5 aggregates shortages by percentage change to base climate delivery for the 2080-Dry scenario. Out of about 343 KAF of total shortage for the 2080-Dry scenario, more than 60% comes from nodes subject to shortages between 5% and 22% of base deliveries. Spatially these nodes are above American Falls, within the Boise/Payette River basins, or in-stream fish flows. This concentration of shortages is the outworking of prior appropriations water rights; that is, it primarily results from the priority dates of junior water rights or from senior water right holders in some locations who lack storage rights and are thus impacted by the earlier spring snowmelt.

The *existing* system capacity can mitigate some long-term average supply shortages (e.g., 2030-Dry), but has greater difficulty coping with long-term average within-year timing shifts (e.g., 2080-Dry). Compared with long-term average supply reductions, within-year timing shortages are easily remediated by additional storage. Serially correlated drought years requires multi-year storage, that is, a five year drought with a 3% average annual supply reduction requires 15% additional storage carried forward from a non-drought period. In contrast, a 3% within-year timing shortage can be resolved with 3% additional storage, even if the timing shortage recurs annually. Nevertheless, additional storage would help with both inter- and intra-year storage problems.

The majority of water deliveries are made with minimal shortages. In the worst case scenario (2080-Dry) about 8.58 MAF, or nearly 70% of base deliveries, are delivered with shortages ranging between 0% and 5% (Figure 4; 2080-Dry cumulative base deliveries between 2 and 10.6 MAF). The cumulative shortage for the 0% to 5% group amounts to 132 KAF, an average shortage of 1.53% (Figure 5; net shortage to the right of -5%). Another 1.85 MAF (15%) of base deliveries are made with no shortages whatsoever, or even small surpluses (Figure 4; deliveries above 10.6 MAF). Shortages for the other climate scenarios are even more benign.

The in-stream fish flows hold a single, relatively junior water right. This flow augmentation granted by the Idaho legislature was not an absolute constraint, but was issued similar to any other water right, with the granting date determining priority. Accordingly, in-stream fish flows are impacted more significantly by the reduction in water supply associated with the 2030-Dry and 2080-Dry scenarios (-6.85% and -9.08%, respectively), than are other reaches of the river system, which include a range of water rights priority dates (Table 3, Section C). The reduction in deliveries below King Hill for the Dry scenarios is due to generally junior water rights held by users in this reach. However, the King Hill reduction is less than for the fish flows because the priority date is senior to that of the fish flows.

In contrast to the fish flow water right, the reaches are aggregates of many users with varying priorities of water rights. Shortages are thus ameliorated or exacerbated depending on seniority, with some exceptions. For example, above American Falls, water rights are relatively senior. Never the less, under the Dry scenarios shortages appear. Despite the seniority in this reach, there is limited storage and much of the existing storage (e.g., the Palisades Reservoir), belongs to junior water rights holders. System wide, even with earlier spring runoff of the two dry scenarios, all or most of the water rights can be satisfied. The worst-case scenario (2080-Dry) averages a 2.73% basin-wide shortage (Table 2 and 3). However, the upstream senior water rights holders cannot store the high springtime flows, so that water either flows downstream or is diverted as storage by junior water right holders downstream who have storage rights in Palisades or other reservoirs further downstream. As a result, water right holders with lower priority are allocated water inaccessible to water rights holders with greater seniority. Consequently, not only are the average shortages above American Falls the highest among the four river segments (Table 3, section C), but some of the individual nodes have shortages exceeding 16% to18% for the 2080-Dry scenario; in fact, 75% of the total shortage for this reach is borne by fewer than 30% of the nodes (MODSIM detail not shown).

The Boise and Payette River basins are separate tributary drainages that join the Snake River on the western border of Idaho. Similar to the Snake River, timing of flows and limited storage are the causes of the average negative impacts, which occur only for the two Dry climate change scenarios, and water rights tend to concentrate these impacts. For the 2080-Dry scenario, one node experiences nearly a 22% shortage and approximately 50% of the total shortages in the Boise and Payette basins are borne by nodes accounting for only 11% of their collective deliveries (MODSIM detail not shown).

4. Summary and conclusions

This study evaluated climate scenario induced changes in irrigation water supply and the impact on deliveries constrained by prior appropriations water rights allocations for Idaho's Snake River Plain. Six climate scenarios were selected to span the spectrum of precipitation of the 90th ('Wet'), median ('Mid') and 10th quintiles ('Dry') for climate change in two periods (2030 and 2080). Intra- and inter-annual variability was incorporated by superimposing an historical, 20-year, daily time step base period onto the climate scenarios. The range of inter-annual variability on water supply is more than an order of magnitude greater than the change imposed by any of the climate scenarios. However, the purpose of this study is to examine the long-term average impact of climate change, so inter-annual variability was filtered out by averaging.

Average annual changes to water supply from each scenario range from a high of +19.4% (2080-Wet) to a low of -5.4% (2030-Dry). For the Mid and Wet scenarios with increased volumes of supply, the impact of earlier flows is largely mitigated. For reduced volumes of flow with limited reductions in summer flow, as in the 2030-Dry scenario, the system partially mitigates the reduced supply volume. For reduced volumes with a shift to earlier spring runoff and lower summer flows, as in the 2080-Dry scenario, the timing shift's interaction with water rights and storage amplifies the average supply shortages into slightly greater average delivery shortages (Table 2). For both the Dry scenarios, the small shortage is concentrated among relatively few nodes. This is primarily the outworking of water rights. The positive side of this concentration is that it maintains a relatively secure and stable water supply for a larger number of users, those with more senior water rights.

Irrigation water shortages are infrequent and small on average. Even in the worst case 2080-Dry scenario, according to MODSIM, average water shortages would range from a low of 0.9% in the American Falls to King Hill region, up to 3.6% above American Falls with a system-wide average shortage of 2.7% relative to base climate delivery. The irrigation system in southern Idaho is very robust in its ability to handle the wide range of climate scenario deficits.

The shortages simulated from climate scenarios result from a complex interaction of the timing and magnitude of runoff, the spatial distribution and seniority of water rights, and the ability to store and redistribute water throughout the system. In some cases, the prior appropriations exacerbate shortages, but more significantly, insufficient storage fails to take advantage of surpluses. Adaptation strategies, such as increased surface or groundwater storage or water transfers, should be sought to address foregone benefits from surpluses. Any added storage would also help reduce negative impacts of shortages.

Other considerations should include conjunctive use management of surface- and ground-water interactions, crop evapotranspiration demand changes associated with climate change, potential long term economic impact from reduced agricultural production associated with climate change due to either supply or demand changes, and land use changes. These are beyond the scope of the present study but will be important.

Long term climate change will necessitate changing intrastate adjudication or interstate compacts. Our study examines shifts in volume and timing of flows and allocations, not the amplitude of inter-year variations, for scenarios of long-term climate change. The prior appropriations doctrine was designed to allocate stochastic variations in inter-year water supplies. Changes in long term averages were not accounted for in setting priority or adjudication. Inter-state compacts, treaties, or other agreements, however, are based upon flows for a base period. The Colorado River compact is a poster child for a compact that was set for a period of wet years and the compact fails during long periods of drought. Our results show that even in the most draconian climate change scenario average shortages in southern Idaho are minor and the doctrine of prior appropriations is capable of managing them. On the other hand, the present system is incapable of taking advantage of water supply surpluses, which the range of climate scenarios indicate are plausible. This is an opportunity worth exploring.

References

Dey, B., Sharma, V.K., & Rango, A. (1989). A test of snowmelt-runoff model for a major river basin in western Himalayas. *Nordic Hydrology, 20*, 167–178.

Eborn, B., Patterson, P., & Taylor, G. (2012). *Financial condition of Idaho agriculture 2012* (Annual Financial report No. 10). University of Idaho Extension, Moscow, Idaho, USA.

Gillan, B., Harper, J., & Moore, J. (2010). Timing of present and future snowmelt from high elevations in northwest Montana. *Water Resources Research, 46*, 1–13, W01507. doi:10.1029/2009WR007861

Gomez-Landesa, E. & Rango, A. (2002). Operational snowmelt runoff forecasting in the Spanish Pyrenees using the snowmelt runoff model. *Hydrological Processes, 16*, 1583–1591. DOI: 10.1002/hyp.1022.

Hall, D.K., Riggs, G.A., & Salomonson, V.V. (2006). *MODIS/Terr540a Snow Cover 8-Day L3 Global 500m Grid V005, January to August, 2003–2006* [Digital Media, updated weekly]. Boulder, CO: National Snow and Ice Data Center. Retrieved from: http://nsidc.org/data/snow. html#SNOW_COVER

Hamilton, J.R., Green, G.P., & Holland, D. (1999). Modeling the reallocation of Snake River water for endangered salmon. *American Journal of Agricultural Economics, 81*, 1252–1256.

Harou, J., Pulido-Velazquez, J.M., Rosenberg, D.E., Medellín-Azuara, J., Lund, J.R., & Howitt, R.E. (2009). Hydro-economic models: Concepts, design, applications, and future prospects. *Journal of Hydrology, 375*, 627–643.

Hong, M.A. & Cheng, G. (2003). A test of snowmelt runoff model (SRM) for the Gongnaisi River basin in western Tianshan Mountains, China. *Chinese Science Bulletin, 48*, 2253–2259. DOI: 10.1360/03wd0135

Houk, E., Frasier, M., & Taylor, R. (2007). Evaluating water transfers from agriculture for reducing critical habitat water shortages in the Platte basin. *Journal of Water Resources Planning and Management, 133*, 320–328.

Intergovernmental Panel on Climate Change (IPCC). (2007). *IPCC fourth assessment report*: Climate Change 2007 (AR4). Cambridge: Cambridge University Press.

Kenny, J.F., Barber, N.L., Hutson, S.S., Lovelace, J.K., & Maupin, M.A. (2009). Estimated use of water in the United States in 2005. *U.S. Geological Survey Circular, 1344*, 52 pp.

Labadie, J.W. (1995). *MODSIM: River basin network flow model for conjunctive stream-aquifer management: Program user manual and documentation*. Ft. Collins, CO: Department of Civil Engineering, Colorado State University.

Martinec, J., Rango, A., & Roberts, R. (2005). *SRM snowmelt runoff model user's manual, updated edition for Windows* (WinSRM Version 1.10, E. Gómez-Landesa, Ed.). Las Cruces, NM: USDA Jornada Experimental Range, New Mexico State University.

Maass, A., Hufschmidt, M., Dorfman Thomas, R., Marglin, H., & Fair, S. (1962). *Design of water-resources systems*. Cambridge, MA: Harvard University Press.

Michelsen, A. & Taylor, R.G. (1999). Endangered species recovery and river basin policy: Contribution of economic analysis. *American Journal of Agricultural Economics, 81*, 1250–1251.

Mote, P., Hamlet, A., & Salathe, E. (2008). Has spring snowpack declined in the Washington Cascades? *Hydrology and Earth System Sciences, 12*, 193–206.

Qualls, R.J. & Arogundade, A.B. (2012). Modeling snowmelt runoff under climate change scenarios using MODIS-based snow cover products. In N.B. Chang & Y. Hong (Eds.), *Multiscale hydrologic remote sensing: Perspectives and applications*. New York, NY: CRC Press, pp. 213–249.

Seaber, P.R., Kapinos, F.P., & Knapp, G.L. (1987). Hydrologic unit maps: U.S. Geological Survey. Retrieved from Water-Supply Paper 2294, 63 p.

Stewart, I.T., Cayan, D.R., & Dettinger, M.D. (2004). Changes in snowmelt runoff timing in western North America under a 'business as usual' climate change scenario. *Climatic Change, 62*, 217–232.

USDA NRCS *Snotel data for Wyoming* (n.d.) [online]. Retrieved from http://www.wcc.nrcs.usda. gov/snotel/Wyoming/wyoming.html

USGS National Water Information System (n.d.) [online]. Retrieved from http://waterdata.usgs.gov/ id/nwis/current/?type=flow?.

US Bureau of Reclamation. *Hydromet network* (n.d.) [online]. Retrieved from: http://www.usbr.gov/ pn-bin/arcread.pl?station=JCK.

Zhang, Y., Li, B., Bao, A., Zhou, C., Chen, X., & Zhang, X. (2007). Study on snowmelt runoff simulation in the Kaidu River basin. *Science in China Series D: Earth Sciences, 50*(Supp. I), 26–35. DOI: 10.1007/s11430-007-5007-4

Climate and choice of irrigation technology: implications for climate adaptation

George B. Frisvold[a] and Shailaja Deva[b]

[a]Department of Agricultural and Resource Economics, University of Arizona; [b]HSBC Bank, Elmhurst

Because studies of irrigation technology adoption often concentrate on small geographic areas with the same climate, few have estimated effects of climate on irrigation technology choice. This study examines the choice of sprinkler versus gravity-flow irrigation across 17 western states. Analysis considers long-term seasonal temperatures and growing season length. An erosion index captures effects of rainfall, field slope, and soil water-holding capacity. Sprinkler adoption increases with reliance on groundwater, pumping costs, farm wage growth, and erosion. Sprinkler adoption was significantly lower for smaller farms. In colder climates, climate warming may lengthen the growing season, but increase susceptibility to frost during the expanded growth period, which may encourage sprinkler adoption. In warmer areas, there is less scope to adapt to warming by switching from gravity to sprinkler technology. Sprinkler adoption declines monotonically in Spring/Summer temperature and growing-season-adjusted Fall/Winter temperature. A drier climate would reduce sprinkler adoption, while climates with more rainfall and more intense rain events would see greater adoption.

Introduction

Improved irrigation efficiency has often been cited as an important way to adapt to climate change (e.g., Burton, 2000; Cavagnaro, Jackson, & Scow, 2006; Jackson et al., 2009; B. Joyce, Mehta, Purkey, Dale, & Hanemann, 2009; Kurukulasuriya & Rosenthal, 2003; Smit & Skinner, 2002). Compared to gravity irrigation, drip or sprinkler irrigation can achieve better control over the timing and level of water applied to crops. This can better match water applications to plant requirements. Improved irrigation timing can help protect crops from drought stress, frost, or other climate extremes. Drip or sprinkler irrigation can require large capital investments, however. Because these capital costs take years to recover, growers have an incentive to select irrigation methods suited to the climate they face. Irrigators often cite financial constraints as major barriers to investing in improved irrigation efficiency (Frisvold & Deva, 2012).

Irrigation technology choice as climate adaptation

Although researchers have thoroughly studied factors affecting adoption of improved irrigation technology, and the importance of climate in irrigation choice is often acknowledged, relatively few studies have formally focused on the role of climate or included climate variables. Many empirical studies of irrigation technology adoption have concentrated on small geographic areas, such as a single irrigation district or relatively small production region. The geographic scope of such analyses can be too narrow to effectively measure effects of climate over the long term. By their very nature, long-term climate averages change little over time, and localized studies may have insufficient variation in climate to allow for econometric analysis. To measure the effects of long-term climate variables, studies must have large enough geographic scope to have measurable differences in these variables across observations. Small geographic scope can also lead to a high degree of multicollinearity between seasonal climate variables, limiting the number of climate variables that can be assessed (Fleischer, Lichtman, & Mendelsohn, 2008).

This study adds to a small list of studies that have focused on the role of climate in adoption of irrigation technology. These include Negri, Gollehon, and Aillery (2005), who examined the effect of climate on the decision whether or not to irrigate; Negri and Brooks (1990), who considered effects of climate variables on choice of irrigation technology among irrigators; Mendelsohn and Dinar (2003), who examined climate effects on the share of irrigated acreage and share of irrigated acreage under different technologies across US counties; and Caswell, Fuglie, Ingram, Jans, and Kascak (2001), who also examined climate effects on the choice of whether to irrigate along with the choice of irrigation method for individual producers. These studies found that climate variables had important impacts on irrigation adoption decisions. In Dinar, Campbell, and Zilberman (1992), temperature and rainfall variables were not significant, but the coefficients of variation for these climate variables were quite low. More recently, Moreno and Sunding (2005) and Schoengold, Sunding, and Moreno (2006) found the number of frost-free days was important for explaining joint crop-irrigation technology choices.

Here, we consider how climate, farm size, water pumping costs, labor costs, and soil characteristics affect irrigator choice between gravity-flow and sprinkler irrigation in 17 western states. To make sure there is enough spatial variation in climate variables, we use data from special cross-tabulations of the Farm and Ranch Irrigation Survey (FRIS) developed by the US Department of Agriculture (USDA)'s Economic Research Service (USDA ERS, 2004). While the USDA conducts the national FRIS roughly every five years, published tables report state-level aggregate data, focusing on 2 x 2 relationships. Thus, one can consider how irrigation technology choices change by farm size *or* pumping costs *or* other factors, but this data configuration is not amenable to multivariate analysis. ERS's Special Tabulation, however, reports data for the 1998 FRIS, stratifying observations by four farm sales classes for the 17 westernmost contiguous US states. This provides 68 state-farm size pairs with sufficient geographic scope to conduct a western region-wide analysis of the choice of irrigation technology.[1]

The study begins with a discussion of the importance of irrigation to western agriculture. Next, we compare the salient features of sprinkler and gravity-flow irrigation technologies and discuss trends in the diffusion of sprinkler technology. We then review the literature on sprinkler irrigation adoption. This literature has examined the role of field and soil characteristics (especially those affecting water-holding capacity of soils), costs of water, labor costs, farm size, and climate as important factors influencing adoption of modern irrigation technologies. The next section introduces a regression specification and data sources to examine how these variables contribute to differences in sprinkler adoption across states and farm sales classes. We then analyze the results of the regression and conclude by discussing the implications of our main findings.

Some main findings follow. Sprinkler adoption rates were significantly lower for farms operating at a smaller scale (measured by sales). An aggregate index of sheet and rill erosion was also a significant predictor of sprinkler adoption. This erosion index embodies variables – rainfall, field slope, and soil water-holding capacity – that have been found to explain sprinkler adoption in farm-level and irrigation district-level studies. Sprinkler adoption increased with water pumping costs and with greater reliance on groundwater. In colder climates, climate warming may lengthen the growing season, but increase susceptibility to frost during the expanded growth period. This may encourage sprinkler adoption for frost protection. In warmer areas, there is less scope to adapt to climate warming by switching from gravity to sprinkler technology. Sprinkler adoption declines monotonically in Spring/Summer temperature and growing-season-adjusted Fall/Winter temperature. The response to the erosion index suggests a drier climate would reduce sprinkler adoption, while climates with more rainfall and more intense rain events would see greater sprinkler adoption.

Western irrigated agriculture

Irrigation is enormously important to western agriculture. In the 17 westernmost contiguous states, about 75% of the value of crops grown come from the 25% of the cropland that is irrigated (Gollehon & Quinby, 2000). Improved irrigation technology has been seen as a necessary, if not sufficient, part of relieving pressure on water supplies and quality in the West. Improved technologies such as sprinkler or drip irrigation allow producers better control over water applications and increase the share of applied water that is taken up by plants. Such improved technologies have been seen as means to improve farm yields and incomes (Aillery & Gollehon, 2003; Caswell & Zilberman, 1986), allow producers to better adapt to drought or climate change (Mendelsohn & Dinar, 2003; Schuck, Frasier, Webb, Ellingson, & Umberger, 2005), and reduce water pollution from soil erosion and chemical leaching (Caswell, Fuglie, et al., 2001; Caswell, Lichtenberg, & Zilberman, 1990; Dressing, 2003).

The role of improved irrigation technology in water conservation has proven more controversial. Although improved application efficiency means less water is needed to generate a given level of crop yield, improved efficiency does not necessarily reduce demand for water (Caswell & Zilberman, 1986; Peterson & Ding, 2005). Even when increased farm-level efficiency does reduce demand for water diversions, it may still increase consumptive use of water. Increasing the percentage of diverted water consumed by crops can reduce return flows and aquifer recharge. Thus, improved farm-level efficiency may not necessarily conserve water at the basin level (Huffaker & Whittlesey, 2000, 2003; Skaggs, 2001; Skaggs & Samani, 2005; Ward & Pulido-Velazquez, 2008).

Irrigation technology

Traditional irrigation systems use gravity to distribute water. Water is conveyed to the field using open ditches or pipes, and then released along the upper end of the field. Furrows control water movement and channel the flow down or across the field. Gravity systems

Table 1.   Shares of irrigated acreage in the US by method of irrigation.

Year	Agency Report	Gravity	Sprinkler	Drip and subsurface
1979	NASS	63%	36%	1%
1995	USGS	52%	40%	3%
1998	NASS	50%	45%	5%
2000	USGS	48%	46%	7%
2003	NASS	43%	51%	6%
2005	USGS	44%	50%	7%
2008	NASS	39%	54%	7%

Sources: USDA NASS (1998, 2004, 2010); Aillery and Gollehon (2003); Solley, Pierce, and Perlman (1998); Hutson et al. (2004); Kenny et al. (2009).

are best suited to soils with higher moisture-holding capacities and relatively flat fields to prevent excessive water runoff.

With sprinkler systems, water is sprayed over the field, usually using aboveground pipes. Sprinkler irrigation systems use pressure to distribute water, which requires energy for pumping. As Negri and Brooks (1990) note, 'Sprinkler irrigation technologies save water relative to gravity-flow systems by distributing water evenly on the field, reducing percolation below the root zone, and eliminating field runoff' (p. 214). Sprinkler systems can be used on steeper slopes unsuited to gravity systems and on soils with lower water-holding capacity. Sprinkler irrigation also can have much higher application efficiencies than traditional gravity irrigation (Sloggett, 1985). While gravity systems have field application efficiencies that usually range from 40% to 65%, efficiencies from sprinkler systems more typically range from 75% to 85% (Aillery & Gollehon, 2003). Sprinkler systems have higher capital costs than gravity systems, however, which may act as a barrier to adoption for farms below some critical size. Sprinkler systems tend to be energy using and labor-saving relative to gravity systems. Thus their relative profitability will depend on labor and energy costs.

The amount of US acreage irrigated with sprinkler systems increased from 36% to 54% between 1979 and 2008. Acreage devoted to drip irrigation rose from 1% to 7%, while acreage irrigated by gravity systems declined from 63% to 39% over the same period (Table 1). Drip irrigation is most commonly used for vegetables, orchards, vineyards, nuts, and other perennial crops. In the 17 Western States, 81% of acreage under drip irrigation is in California, primarily on citrus and specialty crops. While drip (trickle, low-flow and micro-sprinkler) irrigation accounts for 32% of California's irrigated acreage, it accounts for just 1.6% of irrigated acreage in the remaining 16 states (USDA National Agricultural Statistics Service [NASS], 2010). Because of limited adoption of drip systems outside of California, we focus here on the choice between sprinkler and gravity systems.

Table 2 reports acres irrigated with sprinklers (as a proportion of total acres irrigated by sprinkler and gravity methods) for major crops in the West from the most recent (2008) FRIS (USDA NASS, 2010). Table 2 highlights some interesting differences between California and the other 16 Western States. California accounts for a relatively large share of the western US's specialty crop (vegetable, orchard, vineyard, and nut trees) acreage and a small share of acreage of many field crops (wheat, hay, corn, and barley). The proportion of acres irrigated with sprinklers varies by both crop and region. While sprinkler adoption on field crops tends to be relatively low in California, it tends to be high in the remaining 16 Western States. Studies examining joint irrigation

Table 2. Crop specialization and adoption of sprinkler irrigation by crop: California and the rest of the West.

Crops	Percentage of Regional Irrigated Acreage		Percentage of Region's Acreage Irrigated with Sprinklers	
	California	Other 16 Western States	California	Other 16 Western States
Alfalfa	19%	81%	12%	53%
Vegetables	36%	64%	54%	54%
Orchards	43%	57%	14%	26%
Cotton	22%	78%	9%	38%
Wheat	11%	89%	7%	66%
Hay	9%	91%	10%	14%
Potatoes	6%	94%	95%	96%
Corn	3%	97%	1%	72%
Barley	6%	94%	23%	67%

Source: USDA NASS (2010).

technology-crop choices have focused on a single irrigation district in California (e.g., Green, Sunding, Zilberman, & Parker, 1996; Moreno & Sunding, 2005; Schoengold et al., 2006). Table 2, however, suggests that these joint technology-crop choices appear quite different outside of California and suggests some caution is warranted in extrapolating behavior outside of California. An interesting area of future research would be to apply the joint crop-technology framework to a wider geographical area.

Irrigation technology choice: the literature

Caswell and Zilberman (1986) introduced a theoretical model of irrigation technology choice that characterized modern irrigation systems (sprinkler and drip systems) as land quality augmenting. Modern systems, they argued, enhanced the water-holding capacity of soils. Thus improved irrigation technologies would be relatively more profitable to adopt on land with poorer water-holding capacity. Subsequent empirical literature supports Caswell and Zilberman's theoretical specification. This suggests soil and field characteristics matter for choice of irrigation technology and that growers are more likely to adopt modern irrigation technologies for soils with lower water-holding capacity (Dinar & Yaron, 1990; Dinar et al., 1992; Green & Sunding, 1997; Mendelsohn & Dinar, 2003; Negri & Brooks, 1990; Schuck & Green, 2001). Sprinkler adoption rates are generally higher on fields with steeper slopes, which also reduce water-holding capacity.

Economic theory and previous empirical findings suggest there are systematic relationships between scale of operation and irrigation technology choice. Leib, Hattendorf, Elliott, and Matthews (2002) found significant positive relationships between farm size and adoption of scientific irrigation scheduling methods (use of crop evapotranspiration data and soil moisture testing) among Washington farmers. In a study of New Mexico irrigators, Skaggs and Samani (2005) reported a 'lack of interest in making improvements to current irrigation systems or methods on the smallest farms' (p. 43). Comparing irrigation districts in Alberta, Canada Bjornlund, Nicol, and Klein (2009) found evidence of greater adoption of information-intensive irrigation management in areas with larger farms. Frisvold and Deva (2012) found that smaller farms (measured by sales) in the US West were, 'less likely to investigate irrigation improvements, use management-intensive methods for irrigation scheduling, or participate in cost-share programs to encourage adoption of improved irrigation practices' (p. 569).

Sprinkler systems require fixed capital investments, so average fixed costs fall with the scale of operation. One might expect, then, that adoption rates would increase with scale. Evidence of the impact of farm size on sprinkler adoption is mixed, however. Skaggs (2001) and Schuck et al. (2005) found evidence of a positive relationship between farm size and adoption by New Mexico and Colorado irrigators, but Negri and Brooks (1990) reported the opposite result in a large sample of western groundwater-using farms from the 1984 FRIS. Green et al. (1996), Green and Sunding (1997), and Dinar et al. (1992) observed positive relationships between sprinkler adoption and farm field size.

The different ways researchers measure scale of operation complicates comparisons of studies considering farm size and irrigation technology adoption. Some measure scale in terms of sales volume. In addition to areal scale, this may also be capturing the effects of growing high-value specialty crops. Thus, cropping decisions may drive technology choices. Analyses that focus on total farm area may be, in contrast, capturing the effects of low-value, land extensive crops (e.g., irrigated pasture). As a reviewer has noted, there is also a distinction between farm scale and field scale. Technological factors may account for economies of scale at the field level. Scale of farm operation (that includes multiple fields) may affect adoption via relationships between farm operation scale and risk aversion or access to credit.

A number of studies have examined the role of water cost on sprinkler adoption. Caswell and Zilberman (1985), Negri and Brooks (1990), and Dinar et al. (1992) found a positive relationship between water costs and sprinkler adoption. Green et al. (1996) reported that water price had a negative but statistically insignificant effect, whereas Green and Sunding (1997) found water price had a positive effect on sprinkler adoption in citrus production, but not in vineyards.

Empirical findings for the relationship between water source and technology choice have also been mixed. According to Negri and Brooks (1990), greater reliance on surface water decreased the probability of sprinkler adoption. Moreno and Sunding (2005) found such a negative relationship to be statistically significant. Caswell and Zilberman (1985), Dinar et al. (1992), and Green et al. (1996) found a negative, but statistically insignificant relationship between reliance on surface water and sprinkler adoption. Mendelsohn and Dinar (2003), in contrast, found a significant, positive relationship between reliance on surface water and sprinkler adoption.

The number of studies on the effects of climate variables is limited because smaller-scale studies at, for example, the irrigation district level may not exhibit sufficient cross-sectional variation in climate. Studies covering wider geographic areas, however, have found climate to have important effects on irrigation technology choice. Both Negri and Brooks (1990) and Mendelsohn and Dinar (2003) found higher rates of sprinkler adoption in areas with greater rainfall. Dinar et al. (1992) did not find a significant rainfall effect, but this analysis was confined to the San Joaquin Valley of California, which has less variation in rainfall than the other two studies. Negri and Brooks argued that sprinkler adoption would be greater in areas with greater rainfall. This is because growers in high-rainfall areas face a greater risk of crop damage from unexpected rainfall following flood irrigation; here sprinklers provide growers with greater control over applications. In contrast, in hot, arid regions, evaporation losses are large with sprinkler technology. Evaporation losses in sprinkler systems can reach levels close to 50% under the hot, arid conditions found in Arizona or Southern California (McLean, Sri Ranjan, & Klassen, 2000). Negri and Hanchar (1989) state, 'Farmers in hot or windy regions are more likely to adopt gravity since a large fraction of water applied with sprinkler systems evaporates under these climate conditions' (p. 9). Based on farm-level analysis of California, Oregon, and Washington irrigators, Olen, Wu, and Langpap (2012) argue that above a critical temperature threshold, high evaporative losses from sprinklers negate water application efficiency advantages of sprinklers, making them less attractive relative to gravity or drip systems. Mendelsohn and Dinar (2003) also pointed out the problem of large evaporation losses and noted 'sprinkler systems are more frequently adopted in cooler locations with a lot more rainfall' (p. 338).

Negri and Brooks (1990) reported that sprinkler adoption was lower in areas with more frost-free days and a longer growing season (measured in growing degree days) and greater in areas with more frost. They argued that sprinklers are better suited for irrigation for frost protection, whereas longer growing seasons are associated with warmer climates, where sprinkler evaporation losses are greater. Olen et al. (2012) cite several studies reporting the frost protection advantages of sprinkler irrigation over other systems. Mendelsohn and Dinar (2003) also pointed out that sprinkler adoption rates were inversely related to temperature. Dinar et al. (1992) again found no significant effect, but as before, this may be because of the low variance of weather and climate variables in their sample.

Finally, because sprinkler systems tend to be laborsaving, some studies have considered impacts of labor costs on sprinkler adoption. Negri and Brooks (1990) found that higher farm labor wages were associated with greater adoption of sprinkler irrigation systems. Mendelsohn and Dinar (2003) reported greater use of sprinkler systems in counties with higher farm wage rates in January, but lower wage rates in April.

Data and regression model specification

Irrigation data come from special cross-tabulations of the 1998 FRIS (USDA NASS, 1998) made available by USDA's ERS (USDA ERS 2004). Although the regular FRIS report does not report detailed data by farm sales class, the ERS special cross-tabs report data for each of the 17 westernmost contiguous states by four farm sales classes:

- Small farms, with sales less than $100,000;
- Medium farms, with sales from $100,000 to $249,999;
- Large farms, with sales from $250,000 to $499,999; and
- Very large farms, with sales of $500,000 or greater.

The 17 states in the database are North Dakota, South Dakota, Nebraska, Kansas, Oklahoma, Texas, Montana, Wyoming, Colorado, New Mexico, Idaho, Utah, Arizona, Nevada, Washington, Oregon, and California.

Let the proportion of acres irrigated with sprinkler systems (as a share of acreage under sprinkler plus gravity irrigation) by irrigators in sales class i and state j be PS_{ij} while the proportion of acres irrigated with gravity systems be PG_{ij}. Drip and subsurface irrigation acreage were not included, as data were not reported for many states and farm sales classes because of insufficient observations.

Following previous work on irrigation adoption using proportions data (Caswell & Zilberman, 1985; Mendelsohn & Dinar, 2003; Schaible, Kim, & Whittlesey, 1991), the regression equation explaining adoption is specified as a logistic function:

$$ln(PS_{ij}/(PG_{ij})) = \alpha_0 + \alpha_1 \text{Small} + \alpha_2 \text{Medium} + \alpha_3 \text{Large} + \beta' X_{ij} + u_{ij} \qquad (1)$$

where

Small = 1 for small farms, = 0 otherwise;
Medium = 1 for medium farms, = 0 otherwise;
Large = 1 for large farms, = 0 otherwise;
X_{ij} = a vector of other explanatory variables (discussed below); and
u_{ij} = a stochastic error term.

A regression intercept α_0 is included; so, one farm class dummy (for very large farms) is omitted. Coefficients for the other sales-class variables represent differences from the very large farm class. Operations may be in the larger sales classes if they have more acreage, grow higher-value crops, or both. The sales class variables may therefore capture some combination of acreage and crop value effects.

The variable *PUMPING COSTS* is the average energy costs (in $ per acre; one acre equals 0.404686 hectares) to pump water by farm size and state, for 1998 FRIS irrigated farms. Irrigators use pumps to bring well water to the surface; relift or boost water within irrigation systems; discharge water from ponds, lakes, reservoirs, and rivers; or discharge water from tailwater pits. Pumping costs increase with well depth and energy prices. Water pumping costs are a weighted average, accounting for the proportion of water pumped by different energy sources (electricity, natural gas, diesel, etc.) and coming from different water sources (surface and groundwater).

The variable *SURFACE WATER* is the share of surface water to total irrigation water applied by farm size and state, for 1998 FRIS irrigated farms. Irrigators' relative reliance may affect sprinkler adoption in at least two ways. First, as discussed above, marginal costs of surface water are usually lower than costs for groundwater. Previous research suggests that sprinkler adoption increases with water costs. Second, surface water diversions naturally complement gravity-flow systems.

We also include three climate variables: *LNTEMP5-9*, *LNTEMP10-3*, and *BELOW32*. *LNTEMP5-9* is (the log of) the average of monthly 40-year mean temperatures in degrees Fahrenheit of cropland in a state from May through September. *BELOW32* is the total number of months the 40-year mean monthly temperature is less than 32 degrees F. *LNTEMP10-3* is (the log of) the average of monthly 40-year mean temperatures of cropland from October through March, adjusted for growing season. Winter months with average temperatures below 32 degrees F (0 degrees C) are not included in the *LNTEMP10-3*. Sprinkler irrigation is often used to protect crops from frost (Dressing, 2003; Moreno & Sunding, 2005; Negri & Brooks, 1990; Olen et al., 2012; Skaggs, 2001). *LNTEMP10-3* is intended to measure such frost risk. Temperatures in months of freezing temperatures represent the off-season and are not likely to affect this decision. Above freezing, but low Fall/ Winter temperatures suggest areas where frost risk is greater, and sprinkler adoption is higher. One would thus expect an inverse relationship between *LNTEMP10-3* and frost protecting uses of sprinkler irrigation.

The climate variables come from Teigen and Singer (1988), who weighted average weather station measurements by harvested cropland. The climate variables thus give more weight to temperature readings where crops are grown. For example, readings from Death Valley or high in the Rocky Mountains with no agricultural production would receive no weight. In contrast, readings from major agricultural production areas would receive great weight.

The variable *SREROSION* measures average annual sheet and rill erosion on cultivated cropland in each state, measured in tons of soil per acre per year. Data come from the 1997 Natural Resources Inventory of the USDA's Natural Resources Conservation Service

(USDA NRCS 2000). Sheet and rill erosion is caused by water. *SREROSION* is derived from the universal soil loss equation (USLE) (Institute of Water Research, Michigan State University [IWR-MSU], 2002; Wischmeier & Smith, 1978). It is included because it comprises several variables that past studies have found affect irrigation technology choice.

SREROSION = *RKLSCP*, where *R* is a rainfall factor, *K* is a soil erodibility factor, *L* is a slope length factor, *S* is a slope steepness factor, *C* is a cover and management factor, and *P* is a conservation practice factor. *R* is a factor measuring erosivity of soil from rainfall runoff. It increases with the total amount and peak intensity of rainfall. The USLE's *R* factor is highly correlated with measures of precipitation (with $R^2 \sim 0.86$–0.91) (de Santos Loureio & de Azevedo Coutinho, 2001; Diodato, 2004; Yu & Rosewell, 1996). The USLE measure of erosion has been found to be more sensitive to changes in the *R* factor than to other environmental variables (Nearing, 2001).

The factor *K* assigns values to different types of soil based on susceptibility to erosion and rate of runoff. *K* is thus related to the water-holding capacity of soils. Sprinkler irrigation technology enhances the water-holding capacity, and previous studies have found higher adoption rates on soils with low water-holding capacity. For example, clay soils have *K* values ranging from about 0.05 to 0.15, because they resist detachment. Loam soils tend to have moderate *K* values, and sandy soils have higher *K* values. Both Negri and Brooks (1990) and Mendelsohn and Dinar (2003) reported that farming on sandy soils encouraged sprinkler adoption, while farming on clay soils discouraged sprinkler adoption, relative to farming on loam soils. These results also suggest that sprinkler adoption would increase in *SREROSION*, via the relationship with *K*.

Land with steeper slopes has also been associated with greater adoption of sprinkler irrigation relative to gravity irrigation. Mendelsohn and Dinar (2003), however, argued that slope length should be positively associated with gravity irrigation, because it implies flatter fields. They found empirical evidence to support this argument. Slope length appears to be the only variable that increases *SREROSION* and has been associated with less sprinkler irrigation use. All the other components of the USLE appear to both increase *SREROSION* and contribute to greater sprinkler adoption. Thus slope length may have a confounding effect. However, computed values for *SREROSION* are not very sensitive to slope length, particularly on flat landscapes (IWR-MSU, 2002). In sum, all the individual factor components that increase *SREROSION* appear to be positively associated with greater sprinkler adoption except for slope length, *L*. Slope length, however, exerts relatively minor influence on *SREROSION*. Overall, then, one might expect a positive association between *SREROSION* and sprinkler adoption.

Several studies suggest gravity systems are more labor intensive than sprinkler systems (Bernardo, Whittlesey, Saxton, & Bassett, 1987; Maddigan, Chern, & Rizy, 1982; Negri & Brooks, 1990; Negri & Hanchar, 1989; Sauer et al., 2010). Sprinkler systems may thus represent a laborsaving change from gravity systems. Our data compares adoption of irrigation technology at a single point in time. However, such technology decisions often require lumpy capital investments. Growers may not make investments to change irrigation technology based solely on the current year's wages. In preliminary analysis, current wages were positively associated with sprinkler adoption, but the relationship was insignificant. As an alternative to considering only current wages, we estimated the effects of wage *growth rates* in sprinkler system adoption. The variable used *LNWAGE*Δ was the log of 1998 hired farm labor wages minus the log of 1993 wages. This represents the continuous growth rate in hired labor wages between the 1998 and 1993 FRISs. Annual wage rates are reported by multistate region rather than state in the USDA's Quick Stats database (http://quickstats.nass.usda.gov/).

Table 3. Descriptive statistics for variables used in irrigation choice regression.

Variable name	Variable description	Mean	Minimum	Maximum
PS_{ij}	Acreage irrigated with sprinkler systems as a proportion of the sum of area to irrigate with sprinkler and gravity systems	0.49	0.02	0.95
$Ln\ (PS_{ij}\ /\ PG_{ij})$	Ln of the proportion acres irrigated with sprinklers over the proportion of acres irrigated with gravity methods	1.85	0.02	18.01
PUMPING COSTS	Average energy pumping costs ($ per acre) weighted by energy and water source	$34.34	$11.00	$147.12
SURFACE WATER	Proportion of surface water to total irrigation water from all sources	0.59	0.007	0.99
LNTEMP5-9	Ln of average of long-run monthly temperatures from May to September	4.207	4.107	4.413
	Antilog of LNTEMP5-9 (F)	67.2	60.8	82.5
	Antilog of LNTEMP5-9 (C)	19.6	16	28.1
LNTEMP10-3	Ln of growing season adjusted average of long-run month temperatures from October to March	3.783	3.659	4.044
	Antilog of LNTEMP10-3 (F)	43.9	38.8	57.0
	Antilog of LNTEMP10-3 (C)	6.6	3.8	57.0
SREROSION	Average annual sheet and rill erosion on nonfederal land	1.99	0.20	13.9
BELOW 32	Number of months the mean monthly temperature was less than 32 degrees F (0 degrees C)	2.12*	0.00	5.00
LNWAGEΔ	Five-year continuous growth rate of hired farm labor wages	0.073	0.024	0.146

Note: *median $= 2$, mode $= 0$.

Descriptive statistics for the variables are presented in Table 3. Adoption rates for sprinkler irrigation range across state-farm size pairs from 2% to 95%. There are also wide ranges between minimum and maximum values for most of the explanatory variables used in the regression analysis. Growing season adjusted Fall/Winter temperatures range from 39 to 57 degrees F (4 to 14 degrees C), while Spring/Summer temperatures range from around 61 to 83 degrees F (16 to 28 degrees C). The number of months with average temperatures below 32 degrees F (0 degrees C) ranges from zero (which is the mode) to five months, with a median of two months. Table 4 reports differences in water variables by farm sales class. The proportion of acres irrigated with sprinkler systems increases with farm sales class. Reliance on surface water also declines with farm sales class. Water pumping costs are somewhat lower for small and medium operations compared to larger ones.

Sprinkler system adoption: regression results

The regression equation was estimated using ordinary least squares (Table 5). With the log proportions transformation of the dependent variable, we failed to reject the null hypothesis of homoskedasticity. The coefficients for small, medium, and large farms were negative and statistically significant, indicating that these operations had a lower percentage of acreage under sprinkler systems than did very large farms. The coefficient for the variable SMALL (for the smallest farms) was the most negative.

Table 4. Water variables for 17 Western States by farm sales class.

	Small	Medium	Large	Very Large
Percentage of irrigated acres irrigated with sprinklers				
Mean	39%	47%	53%	60%
Min	5%	17%	2%	10%
Max	74%	84%	81%	95%
Water pumping costs ($ / acre)				
Mean	$ 30.19	$ 32.23	$ 39.42	$ 35.54
Min	$ 11.00	$ 13.56	$ 12.70	$ 17.17
Max	$ 98.81	$ 117.53	$ 147.12	$ 54.20
Surface water as a percentage of all water sources				
Mean	66%	60%	56%	53%
Min	5%	6%	2%	1%
Max	100%	94%	97%	97%

Source: USDA ERS (2004).

Table 5. Factors affecting the adoption of sprinkler irrigation ordinary least squares* results.

Dependent variable: $ln\,[(PS_{ij}/PG_{ij})]$ where PS_{ij} is the proportion of acreage irrigated with sprinkler systems and PG_{ij} as a proportion of acreage irrigated with gravity systems.
Adjusted R-squared: 0.80. Number of observations: 68.

Variable	Parameter coefficient	Standard error	P value
Intercept	39.386	7.758	0.0000
SMALL	−0.793	0.206	0.0003
MEDIUM-LARGE	−0.490	0.172	0.0059
PUMPING COSTS	0.011	0.005	0.0157
SURFACE WATER	−1.790	0.428	0.0001
LNTEMP5-9	−6.216	2.272	0.0082
LNTEMP10-3	−3.840	1.399	0.0081
BELOW32	0.071	0.066	0.2872
SREROSION	0.575	0.072	0.0000
*LNWAGE*Δ	14.053	3.631	0.0003

Note: *The null hypothesis of homoskedasticity could not be rejected.

We could not reject the null hypothesis that the regression coefficients for medium and large farms were equal ($\alpha_2 = \alpha_3$) based on an F test (P value = 0.88). We therefore report results with a single variable combining medium and large operations, *MEDIUM-LARGE*. Evaluating marginal effects at regressor sample means (except *BELOW32*, where the median number of months is used), the proportion of acres using sprinkler systems is 0.43 for small operations, 0.51 for medium-large operations, and 0.63 for the largest operations.

While the regression is linear in $ln(PS_{ij}/(PG_{ij}))$ it is non-linear in PS_{ij}, the proportion of irrigated acres irrigated by sprinklers. The predicted value of $PS_{ij} = E[PS_{ij}]$ is

$$E[PS_{ij}] = \{1 + \exp\,[-(\mathbf{X}'_{ij}\boldsymbol{\beta} + \sigma^2/2)]\}^{-1} \tag{2}$$

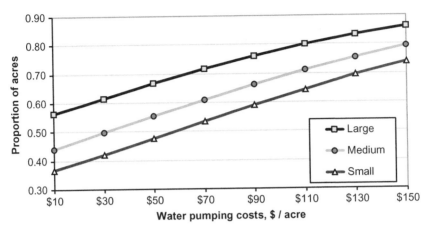

Figure 1. Effect of water pumping costs on the proportion of irrigated acres irrigated with sprinklers by farm sales class.

where β is the vector of regression coefficients, $\mathbf{X}_{ij}$ is a vector of regressors and σ^2 is the variance of the regression equation. The non-linear structure of this logistic function, implies three things: (a) the marginal effect of a change in a variable X_1 depends on values of the other variables; (b) the marginal effect declines as the baseline proportion of adoption moves away from 0.5; and (c) the elasticity of sprinkler adoption with respect to a change in an explanatory variable approaches zero as the underlying proportion of sprinkler adoption approaches one. One can examine how the proportion of acres irrigated by sprinklers changes for changes in continuous variables for different farm sales classes.

In the regression, water-pumping costs had a positive and significant effect on sprinkler adoption (Table 5). This result is consistent with earlier findings suggesting that sprinkler irrigation increases with water costs (Caswell & Zilberman, 1985; Dinar et al. 1992; Green et al., 1996; Moreno & Sunding, 2005; Negri & Brooks, 1990). Figure 1 illustrates the effect of increasing water pumping costs on the proportion of acres irrigated with sprinklers (with other variables set to sample means or medians). The farm sales class variables effectively shift up the adoption curve in Figure 1. Irrigators face a wide range in pumping costs (from $11 / acre to $147 / acre) reflecting differences in well depths and relative reliance on surface and groundwater. For every $20 / acre increase in pumping costs, the proportion of acres irrigating with sprinklers increases about three to five percentage points. The marginal effect declines as pumping costs and adoption rates grow large.

The proportion of acres under sprinkler systems decreased with greater reliance on surface water. Reliance on surface water appears to have a negative effect on sprinkler adoption beyond differences in water pumping costs. Irrigating with surface water entails lower pumping costs than pumping groundwater from wells. Yet, this cost difference is already reflected in the *Water Pumping Cost* variable. Negri and Brooks (1990) note that marginal costs for water from surface sources can be quite low. Surface water supplies in the West are often quantity-rationed rather than price-rationed. In many western surface water projects, average costs can deviate substantially from marginal costs. Indeed, in many cases, irrigators are charged per acre of irrigated land, rather than per acre-foot of water applied. In these cases, the marginal cost of water (per acre-foot) is zero. The fact that sprinkler irrigation adoption rates increase with reliance on groundwater in the regression may also reflect the lower *marginal* cost of surface water. Moreno and Sunding (2005) also note that costs of surface water are less variable than costs of groundwater, which can

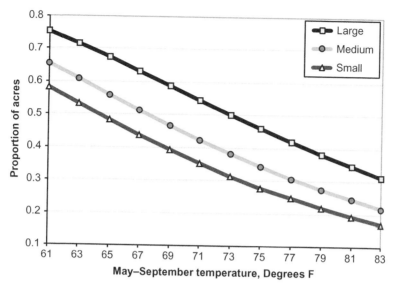

Figure 2. Effect of May–September average temperature on the proportion of irrigated acres irrigated with sprinklers by farm sales class.

fluctuate based on changes in the water table and volatile energy costs. Thus, irrigators relying on groundwater may face both higher and more variable marginal costs of water. Both factors may encourage adoption of more water-efficient technologies.

Turning to the temperature variables, there is a significant negative relationship between May–September temperatures and sprinkler adoption. We experimented with quadratic and log quadratic specifications of this temperature variable. In no case was the quadratic term significant, suggesting that sprinkler adoption is monotonically decreasing in May–September temperatures. Using the same procedure, it was also found that sprinkler adoption was monotonically decreasing in growing-season-adjusted Winter/Fall temperatures (*LNTEMP10-3*).

The proportion of acres using sprinkler irrigation is evaluated at the regressor sample means (median for *BELOW32*) for different ranges of May–September temperatures (Figure 2). At the lower range of observed data (about 61 degrees F), sprinkler adoption is relatively high. As May–September temperatures increase, the expected proportion of sprinkler acres declines. Adoption rates for smaller scale operations is everywhere lower than for larger operations. At the limit of observed data – about 83 degrees F (28 degrees C) – the sprinklers are expected to be used on about 30% of acres on the largest operations and on less than 20% of acreage irrigated by the smallest. A similar pattern holds for growing-season-adjusted Fall/Winter temperatures. As temperatures reach the upper end of observed data, sprinkler adoption rates among acres irrigated by the smallest operations falls to about 20% of irrigated acreage (Figure 3).

These results suggest that, in areas with warmer climates, there may be limited scope to adapt to climate warming via adoption of sprinkler irrigation systems. In cooler climates, the relationship between climate warming and sprinkler adoption is more complex. The variable *LNTEMP10-3* does not include fall and winter months with average temperatures below 32 degrees F (0 degrees C). This is because one would not expect changes in off-season temperatures to affect irrigation decisions for the growing season. However, climate warming may convert an off-season month to a relatively cold growing-season month. If it

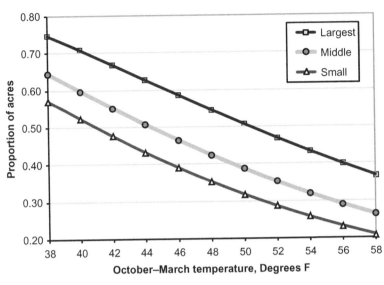

Figure 3. Effect of growing-season-adjusted October–March temperature on the proportion of irrigated acres irrigated with sprinklers by farm sales class.

does so, it is possible that this effect will *lower* the value of *LNTEMP10-3* and encourage sprinkler adoption. This is a mathematical explanation, but not an agronomic one. In agronomic terms, warming in cold climates increases the number of months where plant growth is possible, but simultaneously may increase frost risk in this expanded growth window. Whether climate warming has a positive effect on sprinkler adoption in colder states, however, will depend on a complex combination of changes in the length of growing season, Fall/Winter temperatures and Spring/Summer temperatures. Our results do suggest, however, that in warmer western states (Arizona, California, New Mexico, Oklahoma, Oregon and Texas), climate warming would discourage sprinkler adoption.

These results regarding temperature apply only to the choice of sprinkler irrigation relative to gravity irrigation (i.e., changes at the intensive margin). They say nothing about whether sprinkler adoption may encourage more land to be brought under irrigated cultivation. Caswell and Zilberman (1986) discuss conditions where introduction of sprinklers would increase total acreage under irrigation. Caswell, Fuglie, et al. (2001) found evidence of warmer temperatures increasing sprinkler irrigation adoption, relative to dryland production. There is also the possibility that some irrigators may switch to drip irrigation in response to warmer temperatures (e.g., Mendelsohn & Dinar, 2003; Olen et al., 2012). Other climate adaptations include switching the mix of crops grown and deficit irrigation (Frisvold & Konyar, 2012; Moreno & Sunding, 2005; Olen et al., 2012; Schoengold et al., 2006).

The coefficient on average annual sheet and rill erosion is positive and significant, indicating that in areas with high erosion, sprinkler irrigation systems are more likely to be adopted. Because sheet and rill erosion is more likely on areas with greater rainfall, steeper slopes, and poorer water-holding capacity, it is likely this variable is picking up the effects of combinations of these factors. The results here are consistent with previous research suggesting sprinkler adoption will be greater in areas with greater precipitation where it can supplement rainfall (Mendelsohn & Dinar, 2003; Negri & Brooks, 1990) and on lower quality soils (Caswell & Zilberman, 1985; Mendelsohn & Dinar, 2003; Negri &

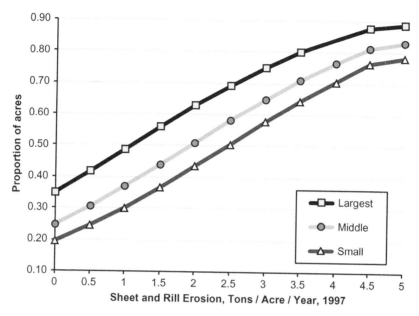

Figure 4.   Effect of state average sheet and rill erosion on the proportion of irrigated acres irrigated with sprinklers by farm sales class.

Brooks, 1990). Figure 4 illustrates how the proportion of acres irrigated with sprinklers increases with a state's index of sheet and rill erosion. In arid regions such as Arizona, California, and Nevada, there is less scope for sheet and rill erosion because there is less rainfall. In these states, the index is near its lowest level among observations (0.2–0.7). In this range, the proportion of acreage irrigated with sprinklers would be expected to be in the 20% to 30% ranges among the smallest scale operators.

Finally, because sprinkler systems tend to be laborsaving, we would expect that higher wages would encourage sprinkler adoption. Negri and Brooks (1990), using annual wage rates, find a positive relationship. The regression coefficient for wage growth is positive and significant in our analysis (Table 5). Figure 5 shows the relationship between growth rates of farm wages and sprinkler adoption for different farm sales classes. At low rates of wage growth, the marginal effect of wage growth on sprinkler adoption is largest among the largest scale operators (i.e., the slope of the curve in Figure 5 is steeper). One might expect larger operations that rely more on hired labor would be more sensitive to wage changes. However, the rate of adoption decreases as adoption rates approach 100%. For the largest operations, once sprinkler adoption exceeds 70%, wage growth exerts a noticeably weaker effect on adoption.

Policy implications and conclusions

This study illustrates how special tabulations of the USDA FRIS can be used for multivariate regression analysis. A simple model that divides irrigators into state-farm size pairs explains 80% of the variation in the extent of adoption of sprinkler irrigation relative to gravity irrigation across 17 western states. The proportion of acres irrigated with sprinkler irrigation was greatest for the largest operations and lowest among the smallest (in terms of sales). Adoption of sprinkler irrigation was also positively influenced by the extent of sheet and rill erosion, which captures effects of greater rainfall, steeper slopes, and soils

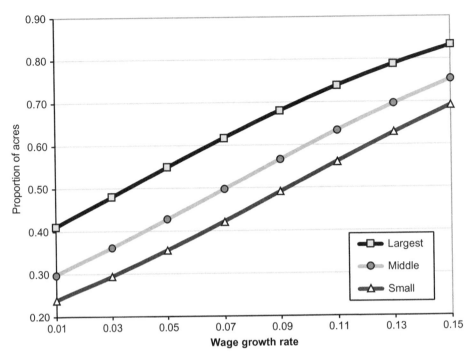

Figure 5. Effect of wage rate growth on the proportion of irrigated acres irrigated with sprinklers by farm sales class.

with less water-holding capacity. Sprinkler adoption increased with water pumping costs, reliance on groundwater and farm wage growth.

Several commentators have suggested adoption of improved irrigation will be important for agricultural adaptation to climate change (e.g., Burton, 2000; Cavagnaro et al., 2006; Jackson et al., 2009; B. Joyce et al., 2009; Kurukulasuriya & Rosenthal, 2003; Smit & Skinner, 2002). A number of studies have raised questions about whether smaller-scale agricultural operators will have the financial or technical capacity to adequately adapt to climate change (L. A. Joyce, Ojima, Seielstad, Harriss, & Lackett, 2001; McLeman, Mayo, Strebeck, & Smit, 2008; Walthall et al., 2012; Wolfe et al., 2008). Our findings suggest smaller scale producers in the western US adopt more efficient sprinkler irrigation systems to a significantly lower extent than their larger-scale counterparts.

The lower rates of sprinkler adoption among smaller-scale producers raises the question of whether policy interventions are warranted to encourage greater adoption by these producers. The Environmental Quality Incentives Program (EQIP) provides cost share payments to US farmers, subsidizing adoption of a variety of conservation practices and investments. Sprinkler adoption is the third most common EQIP-subsidized practice, with growers receiving more than $400 million in payments from 1997 to 2010 (Wallander, Aillery, Hellerstein, & Hand, 2013). Farmers with lower sales have participated less in EQIP and other water- or energy-conservation programs, however (Frisvold & Deva, 2012; Nickerson & Hand, 2009). The 2002 and 2008 Farm Bills included provisions to increase use of EQIP among beginning, socially disadvantaged, or limited resource farmers (Nickerson & Hand, 2009). These groups include farmers with relatively lower sales (our definition of small-scale operations). Schaible (2004) raises the point that there may be an efficiency-equity trade-off in targeting smaller scale irrigators for greater conservation program participation. He points out that larger scale operators account for the bulk of irrigated acreage and irrigation water use. Targeting larger-scale producers may thus lead to greater overall improvements in irrigation technology and water conservation.

This study's results suggest that even among larger scale operations, adoption of sprinkler irrigation may not be a preferred adaptation to climate warming. Sprinkler adoption declined monotonically in Spring/Summer temperature and growing-season-adjusted Fall/Winter temperature. In colder states, complex interactions of changing growing season length and frost risk may encourage sprinkler adoption. For warmer climates, however, warmer temperatures discourage sprinkler adoption. Adoption of sprinkler irrigation was also positively related to an index of sheet and rill erosion, which may proxy for effects of greater rainfall, steeper slopes, and soils with less water-holding capacity. Values for the erosion index have been found to be more sensitive to changes in rainfall amounts and intensity than changes in other environmental variables (Nearing, 2001). Projections of future rainfall for the western US remain highly variable. Nearing (2001) projected that the rainfall component of the erosivity index would increase in some parts of the West under climate change, but decrease in other parts.

The Third National Climate Assessment of the US Global Change Research Program projects a decrease in spring and winter precipitation in the Southwest, which would reduce the erosion index (Walsh & Wuebbles, 2013), which our regression results suggest would discourage sprinkler adoption. However, the Assessment also projects an increase in the frequency and intensity of extreme precipitation events (Walsh & Wuebbles, 2013), which would have the opposite effect. In sum, our results suggest sprinkler irrigation is more likely to be a potential climate adaptation in areas that are relatively cold and where extreme precipitation events increase. Sprinkler adoption is less likely in warmer climates and under drier climate change scenarios. An implication for climate adaptation policy is that for hot, arid parts of the US West, other adaptation strategies – such as deficit irrigation, adoption of drip irrigation, and use of advanced water management practices (such as soil- or plant-moisture sensing devices, commercial irrigation scheduling services, or computer-based crop-growth simulation models) – may be more important to pursue.

Regression results suggest that sprinkler adoption is lower where water-pumping costs are low and where reliance on surface water is greater. One policy option to encourage greater sprinkler adoption may be incentive pricing for surface water. Many irrigation districts do not charge irrigators the marginal cost of surface water, but apply quantity rationing, price water for district infrastructure cost-recovery, or follow other allocation schemes that do not reflect the true scarcity of water. Movements to marginal cost pricing, perhaps under tiered pricing schemes may act to encourage adoption of more efficient irrigation technologies such as sprinklers and drip systems.

Major policy proposals to mitigate climate change usually involve some form of carbon tax or cap-and-trade scheme for emission permits. Both policies would increase the cost of fossil fuels. Such policies can greatly increase costs of groundwater pumping and the relative profitability of growing different crops (Frisvold & Konyar, 2013). Carbon taxes or cap-and-trade policies could have complex implications for irrigation technology. While our results suggest that irrigators shift away from gravity systems under higher pumping costs, sprinkler systems tend to be energy intensive. More research is needed in the area of the role of energy costs, irrigation technology choice, crop choice, and water demand.

The FRIS collects the most detailed, comprehensive data on irrigation practices and water use at the national and state level in the US. Yet, the survey data is seldom used for statistical analysis of irrigator behavior. This is because USDA reports data as tables of state-level aggregates, providing far too few observations for multivariate analysis. To protect respondent confidentiality, access to farm-level FRIS data is restricted. For this reason, rigorous research using FRIS data has been limited to a few studies conducted by USDA economists (and collaborators) (e.g., Negri & Brooks, 1990). A notable exception to this rule is a recent, interesting study by Olen et al. (2012) that analyzed farm-level FRIS data for more than 1000 irrigators in California, Oregon, and Washington. This study has sufficient geographic scope to assess effects of climate and irrigator behavior in detail.

One goal of this article is to illustrate that data such as the Special Tabulation of the FRIS provided by USDA's ERS is a valuable type intermediate data product for researchers without access to farm-level FRIS data. If USDA made more such special tabulations available, then researchers would actually use FRIS data for more than simple descriptions of general irrigation trends and patterns. Another possibility for using publicly available FRIS data could be to assess joint crop-irrigation technology choices in a framework similar to Moreno and Sunding (2005), Schoengold et al. (2006), and Olen et al. (2012), albeit for more aggregate data. The FRIS does report irrigation technology choice, acreage, and water use by crop and state for multiple years. It is possible to exploit the time-series cross-section nature of the data to evaluate a wide geographic area with a sufficient sample size. The geographic scope of the FRIS means it is a valuable source of detailed data to assess climate-irrigation relationships. More and better use of this data would significantly increase our understanding of such relationships.

Acknowledgments

We thank two anonymous reviewers for comments that greatly improved the manuscript. This work was supported by the National Oceanic and Atmospheric Administration's Climate Program Office through Grant NA07OAR4310382 with the Climate Assessment for the Southwest program at The University of Arizona.

Note

1. One reviewer noted that irrigation technology and crop choice can be joint decisions as modeled by Green et al. (1996), Moreno and Sunding (2005), and Schoengold et al. (2006). Thus, explanatory variables may affect results via their unobserved effects on crop choice. The ERS Special Tabulation, however, does not report irrigation technology choice by crop choice. We consider the scope for using aggregate FRIS data from multiple survey years for analysis of joint crop-technology choices in the concluding section of this article.

References

Aillery, M. & Gollehon, N. (2003). Irrigation water management. In R. Heimlich (Ed.), *Agricultural resources and environmental indicators, 2003, agriculture handbook No. AH722* (pp. 134–143). Washington, DC: US Department of Agriculture, Economic Research Service.

Bernardo, D. J., Whittlesey, N. K., Saxton, K. E., & Bassett, D. L. (1987). An irrigation model for management of limited water supplies. *Western Journal of Agricultural Economics, 12*, 164–173.

Bjornlund, H., Nicol, L., & Klein, K. (2009). The adoption of improved irrigation technology and management practices: A study of two irrigation districts in Alberta, Canada. *Agricultural Water Management, 96*, 121–131.

Burton, I. (2000). Adaptation to climate change and variability in the context of sustainable development. In L. Gómez-Echeverri (Ed.), *Climate change and development* (pp. 153–173). New Haven, CT, & New York: Yale School of Forestry and Environmental Studies and United Nations Development Programme.

Caswell, M., Fuglie, K., Ingram, C., Jans, S., & Kascak, C. (2001). *Adoption of agricultural production practices: Lessons learned from the U.S. Department of Agriculture Area studies project*

(Agricultural Economic Report No. AER792). Washington, DC: US Department of Agriculture, Economic Research Service.

Caswell, M., Lichtenberg, E., & Zilberman, D. (1990). The effects of pricing policies on water conservation and drainage. *American Journal of Agricultural Economics, 72*, 883–890.

Caswell, M. & Zilberman, D. (1985). The choices of irrigation technologies in California. *American Journal of Agricultural Economics, 67*, 224–324.

Caswell, M. & Zilberman, D. (1986). The effects of well depth and land quality on the choice of irrigation technology. *American Journal of Agricultural Economics, 68*, 798–811.

Cavagnaro, T., Jackson, L., & Scow, K. (2006). *Climate change: Challenges and solutions for California agricultural landscapes* (White Paper CEC-500-2005-189-SF). California Climate Change Center. Retrieved from www.energy.ca.gov/2005publications/CEC-500-2005-189/CEC-500-2005-189-SF.PDF

de Santos Loureio, N. & de Azevedo Coutinho, M. (2001). A new approach to estimate the RUSLE EI_{30} index based on monthly rainfall data and applied to the Algarve Region, Portugal. *Journal of Hydrology, 250*, 2–18.

Diodato, N. (2004). Estimating RUSLE's rainfall factor in the part of Italy with a Mediterranean rainfall regime. *Hydrology and Earth System Sciences, 8*, 103–107.

Dinar, A., Campbell, M., & Zilberman, D. (1992). Adoption of improved irrigation and drainage reduction technologies under limiting environmental conditions. *Environmental and Resource Economics, 2*, 373–398.

Dinar, A. & Yaron, D. (1990). Influence of quality and scarcity of inputs on the adoption of modern irrigation technologies. *Western Journal of Agricultural Economics, 15*, 224–233.

Dressing, S. A. (2003). *National management measures to control non-point source pollution from agriculture* (EPA- 841-B-03-004). Washington, DC: US Environmental Protection Agency, Office of Water.

Fleischer, A., Lichtman, I., & Mendelsohn, R. (2008). Climate change, irrigation, and Israeli agriculture: Will warming be harmful? *Ecological Economics, 67*, 109–116.

Frisvold, G. B. & Deva, S. (2012). Farm size, irrigation practices, and conservation program participation in the US Southwest. *Irrigation and Drainage, 61*, 569–582.

Frisvold, G. & Konyar, K. (2012). Less water: How will agriculture in Southern Mountain States adapt? *Water Resources Research, 48*,W05534. doi:10.1029/2011WR011057

Frisvold, G. & Konyar, K. (2013). Climate change mitigation policies: Implications for Agriculture and Water Resources. Journal of Contemporary Water Research & Education, 151, (in press).

Gollehon, N. & Quinby, W. (2000). Irrigation in the American West: Area, water and economic activity. *International Journal of Water Resources Development, 16*, 187–195.

Green, G. & Sunding, D. (1997). Land allocation, soil quality and irrigation technology choices. *Journal of Agricultural and Resource Economics, 27*, 267–275.

Green, G., Sunding, D., Zilberman, D., & Parker, D. (1996). Explaining irrigation technology choices: A microparameter approach. *American Journal of Agricultural Economics, 78*, 1064–1072.

Huffaker, R. & Whittlesey, N. (2000). The allocative efficiency and conservation potential of water laws encouraging investments in on-farm irrigation technology. *Agricultural Economics, 24*, 47–60.

Huffaker, R. & Whittlesey, N. (2003). A theoretical analysis of economic incentive policies encouraging agricultural water conservation. *International Journal of Water Resource Development, 19*, 37–55.

Hutson, S.S., Barber, N.L., Kenny, J.F., Linsey, K.S., Lumia, D.S., & Maupin, M.A. (2004). *Estimated use of water in the United States in 2000: USGS Circular 1268.* Reston, VA: US Geological Survey.

Institute of Water Research, Michigan State University (IWR MSU). (2002). *RUSLE online soil erosion assessment tool.* Retrieved from www.iwr.msu.edu/rusle/

Jackson, L., Santos-Martin, F., Hollander, A., Horwath, W., Howitt, R., Kramer, J., . . . Wheeler, S. (2009). *Potential for adaptation to climate change in an agricultural landscape in the Central Valley of California. Final Paper CEC-500-2009-044-F.* Sacramento, CA: California Climate Change Center. Retrieved from www.energy.ca.gov/2009publications/CEC-500-2009-044/CEC-500-2009-044-F.PDF

Joyce, B., Mehta, V., Purkey, D., Dale, L., & Hanemann, M. (2009). *Climate change impacts on water supply and agricultural water management in California's Western San Joaquin Valley, and potential adaptation strategies. Final Paper CEC-500-2009-051-F.* Sacramento, CA: California

Climate Change Center. Retrieved from www.energy.ca.gov/2009publications/CEC-500-2009-051/CEC-500-2009-051-F.PDF

Joyce, L.A., Ojima, D., Seielstad, G., Harriss, R., & Lackett, J. (2001). Potential consequences of climate variability and change for the Great Plains. In National Assessment Synthesis Team (Ed.), *Climate change impacts on the United States: The potential consequences of climate variability and change* (pp. 191–217). Cambridge: Cambridge University Press.

Kenny, J. F., Barber, N. L., Hutson, S. S., Linsey, K. S., Lovelace, J. K., & Maupin, M. A. (2009). *Estimated use of water in the United States in 2005: USGS Circular 1344.* Reston, VA: US Geological Survey.

Kurukulasuriya, P. & Rosenthal, S. (2003). *Climate change and agriculture: A review of impacts and adaptations* (Climate Change Series No. 91). Washington, DC: World Bank.

Leib, B., Hattendorf, M., Elliott, T., & Matthews, G. (2002). Adoption and adaptation of scientific irrigation scheduling: Trends from Washington, USA as of 1998. *Agricultural Water Management, 55*, 105–120.

Maddigan, R. J., Chern, W., & Rizy, C. G. (1982). The irrigation demand for electricity. *American Journal of Agricultural Economics, 64*, 673–680.

McLean, R. K., Sri Ranjan, R., & Klassen, G. (2000). Spray evaporation losses from sprinkler irrigation systems. *Canadian Agricultural Engineering, 42*, 1–8.

McLeman, R., Mayo, D., Strebeck, E., & Smit, B. (2008). Drought adaptation in rural eastern Oklahoma in the 1930s: Lessons for climate change adaptation research. *Mitigation and Adaptation Strategies for Global Change, 13*, 379–400.

Mendelsohn, R. & Dinar, A. (2003). Climate, water, and agriculture. *Land Economics, 79*, 328–341.

Moreno, G. & Sunding, D. (2005). Joint estimation of technology adoption and land allocation with implications for the design of conservation policy. *American Journal of Agricultural Economics, 87*, 1009–1019.

Nearing, M.A. (2001). Potential changes in rainfall erosivity in the U.S. with climate change during the 21st Century. *Journal of Soil and Water Conservation, 56*, 229–232.

Negri, D. H. & Brooks, D. H. (1990). Determinants of irrigation technology choice. *Western Journal of Agricultural Economics, 15*, 213–223.

Negri, D. H., Gollehon, N. R., & Aillery, M. P. (2005). The effects of climatic variability on US irrigation adoption. *Climatic Change, 69*, 299–323.

Negri, D. H. & Hanchar, J. (1989). *Water conservation through irrigation technology. Agriculture information bulletin No. 576.* Washington, DC: US Department of Agriculture, Economic Research Service.

Nickerson, C. & Hand, M. (2009). *Participation in conservation programs by targeted farmers: Beginning, limited-resource, and socially disadvantaged operators' enrollment trends. Economic information bulletin No. 62.* Washington, DC: US Department of Agriculture Economic Research Service.

Olen, B., Wu, J., & Langpap, C. (2012, August 12–14). *Crop-specific irrigation choices for major crops on the West Coast: Water scarcity and climatic determinants.* Selected Paper presented at the Agricultural and Applied Economics Association Annual Meeting, Seattle, WA.

Peterson, J. M. & Ding, Y. (2005). Economic adjustments to groundwater depletion in the High Plains: Do water-saving irrigation systems save water? *American Journal of Agricultural Economics, 87*, 147–159.

Sauer, T., Havlik, P., Schneider U.A., Schmid, E., Kindermann G., & Obersteiner, M. (2010). Agriculture and resource availability in a changing world: The role of irrigation. *Water Resources Research, 46*, W06503.

Schaible, G. (2004). Irrigation, water conservation, and farm size in the Western United States. *Amber Waves, 2*, 8.

Schaible, G., Kim, C., & Whittlesey, N. (1991). Water conservation potential from irrigation technology transitions in the Pacific Northwest. *Western Journal of Agricultural Economics, 16*, 194–206.

Schoengold, K., Sunding, D., & Moreno, G. (2006). Price elasticity reconsidered: Panel estimation of an agricultural water demand dunction. *Water Resource Research, 42*, W09411. doi:10.1029/2005WR004096

Schuck, E. C., Frasier, W. M., Webb, R. S., Ellingson, L. J., & Umberger, W. J. (2005). Adoption of more technically efficient irrigation systems as a drought response. *International Journal of Water Resources Development, 21*, 651–662.

Schuck, E. C. & Green, G. P. (2001). Field attributes, water pricing, and irrigation technology adoption. *Journal of Soil and Water Conservation, 56*, 293–298.

Skaggs, R. K. (2001). Predicting drip irrigation use and adoption in a desert region. *Agricultural Water Management, 51*, 125–142.

Skaggs, R. K. & Samani, Z. (2005). Farm size, irrigation practices, and on-farm irrigation efficiency. *Irrigation and Drainage, 54*, 43–57.

Sloggett, G. (1985). *Energy and U.S. Agricultural Irrigation Pumping, 1974–1983. Agricultural Economics Report No. 545.* Washington, DC: US Department of Agriculture, Economic Research Service.

Smit, B. & Skinner, M.W. (2002). Adaptation options in agriculture to climate change: A typology. *Mitigation and Adaptation Strategies for Global Change, 7*, 85–114.

Solley, W. B., Pierce, R. R., & Perlman, H. A. (1998). *Estimated use of water in the United States in 1995: USGS Circular 1200.* Reston, VA: US Geological Service.

Teigen, L. & Singer, F. (1988). *Weather in U.S. agriculture: Monthly temperature and precipitation by State and farm production region, 1950–86. Statistical bulletin No. 765.* Washington, DC: US Department of Agriculture, Economic Research Service.

US Department of Agriculture, Economic Research Service (USDA ERS). (2004). *Data sets: Western irrigated agriculture: Tables.* Retrieved from http://ers.usda.gov/Data/WesternIrrigation/ShowTables.asp

US Department of Agriculture (USDA NASS). (1998). *Census of agriculture, AC97-SP-1, vol. 3, Special studies, Part 1, 1998 farm and ranch irrigation survey.* Washington, DC: Author.

US Department of Agriculture (USDA NASS). (2004). *Farm and ranch irrigation survey (2003), Vol. 3, Special Studies – Part 1 of the 2002 Census of agriculture, AC-02-SS-1.* Washington, DC: Author.

US Department of Agriculture (USDA NASS). (2010). *Farm and ranch irrigation survey (2008), Vol. 3, Special studies – Part 1 of the 2007 Census of agriculture, AC-07-SS-1.* Washington, DC: Author.

US Department of Agriculture, Natural Resources Conservation Service (USDA NRCS). (2000). *Summary report: 1997 national resources inventory* (Revised December 2000). Washington, DC, & Ames, IA: Natural Resources Conservation Service and Iowa State University Statistical Laboratory.

Wallander, S., Aillery, M., Hellerstein, D., & Hand, M. (2013). *The role of conservation programs in drought risk adaptation, ERR-148.* Washington, DC: US Department of Agriculture, Economic Research Service.

Walsh, J. & Wuebbles, D. (2013). Our changing climate. Chapter 2 in the *Third national climate assessment: Draft report for public comment.* Washington, DC: US Global Change Research Program.

Walthall, C. L., Hatfield, J., Backlund, P., Lengnick, L., Marshall, E., Walsh, M., . . . Ziska, L.H. (2012). *Climate change and agriculture in the United States: Effects and adaptation. USDA technical bulletin 1935.* Washington, DC: US Department of Agriculture, Climate Change Program Office.

Ward, F. A. & Pulido-Velazquez, M. (2008). Water conservation in irrigation can increase water use. *Proceedings of the National Academy of Sciences, 105*, 18215–18220.

Wischmeier, W. H. & Smith, D. D. (1978). *Predicting rainfall erosion losses: A guide to conservation planning. Agricultural handbook No. 537.* Washington, DC: US Department of Agriculture, Soil Conservation Service.

Wolfe, D., Ziska, L. H., Petzoldt, C., Seaman, A., Chase, L., & Hayhoe, K. (2008). Projected change in climate thresholds in the Northeastern US: Implications for crops, pests, livestock, and farmers. *Mitigation and Adaptation Strategies for Global Change, 13*, 555–575.

Yu, B. & Rosewell, C. J. (1996). A robust estimator of the R factor for the universal soil loss equation. *Transactions of the American Society of Agricultural Engineers, 39*, 559–561.

Potential economic impacts of water-use changes in Southwest Kansas

Bill Golden[a] and Jeff Johnson[b]

[a]*Department of Agricultural Economics, 342 Waters Hall, Kansas State University;* [b]*Department of Agricultural and Applied Economics, Texas Tech University*

This research considers three policy scenarios aimed at reducing groundwater consumption in three high priority subareas of southwest Kansas. The three policy scenarios include: (1) a Status Quo scenario where there is no change in water-use policy, (2) an Immediate Conversion to Dryland scenario where all groundwater pumping is halted, and (3) a Reallocation scenario, which allows only a 40% reduction in saturated thickness in 25 years. Each policy scenario is simulated under normal versus drought weather conditions. Results suggest that, from both a community and producer perspective, groundwater conservation policy can generate positive economic gains.

Introduction

Current levels of groundwater consumption in southwest Kansas raise concerns about the long-term feasibility of irrigated agriculture in the area. Over 95% of current groundwater consumption is attributable to agricultural production. In this arid region, irrigated crop production, primarily corn, has given rise to major livestock production and food processing industries. Given current groundwater consumption, portions of the Ogallala Aquifer will become economically exhausted for irrigated agricultural use in the foreseeable future. For all practical purposes, the Ogallala Aquifer is the only source of irrigation in the area; without irrigation, highly productive farmland will begin reverting to dryland production. To extend the aquifer's economic life and maintain the region's economic base, policy intervention may be needed. The purpose of this research is to assist Groundwater Management District #3 (GMD#3) in their groundwater planning and management process. GMD#3 encompasses 12 counties (Finney, Ford, Grant, Gray, Hamilton, Haskell, Kearny, Meade, Morton, Seward, Stanton, and Stevens) in southwest Kansas. Future long-term predictions of groundwater use, hydrological parameters, and various economic indicators were constructed to provide input into the planning process.

This paper summarizes a full report provided to GMD#3. We consider three policy scenarios for three high priority subareas within GMD#3 (Figure 1). The three policy scenarios include: (1) a Status Quo scenario where there is no change in groundwater-use policy, (2) an Immediate Conversion to Dryland scenario where all groundwater pumping

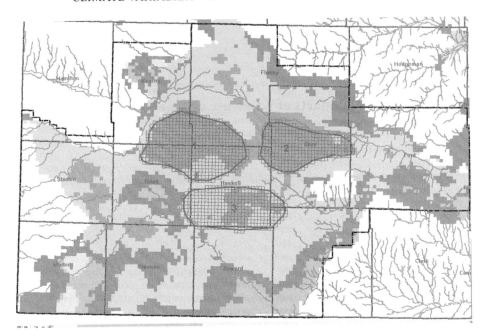

Figure 1. (Color online) Subareas of southwest Kansas.

is immediately halted, and (3) a Reallocation scenario which insures no more than a 40% reduction in current saturated thickness in the next 25 years, recalibrated every 10 years. We use policy scenario 2, Immediate Conversion to Dryland, to value groundwater. Our main focus, however, is to compare policy scenarios 1 and 3. We estimate the three policy scenarios' impacts for both normal and drought weather conditions. As a result, the modeling of each subarea consists of six future scenarios. Due to space limitations, only results for Subarea 1 are discussed in detail. However, relevant results for Subareas 2 and 3 are briefly summarized.

GMD#3 and the Kansas Division of Water Resources identified these three subareas as high priority because they overlay subunits of the Ogallala Aquifer that are in rapid decline. Hydrological data from the Kansas Geological Survey were used to determine the aquifer subunits' boundaries. Water-use goals were then established to extend and conserve the life of the Ogallala Aquifer.

Model overview

The research described in this paper was the second phase of a larger project funded by the Kansas Water Office and GMD#3. During the first phase of that project, the Kansas Geological Survey (KGS) constructed a groundwater flow model and applied the model to various scenarios selected by GMD#3 and the Kansas Water Office. The economic portion of the project required the development of two economic models: a 'temporal allocation' model, and a 'regional economic impact' model. The temporal allocation model provides time-series forecasts of groundwater use, irrigated acreage, and economic productivity within each subarea, and for every policy-weather combination or scenario. The regional economic impact model uses output from the temporal allocation model to predict the impacts of each policy-weather scenario on the subareas' rural economies. This modeling process is illustrated in Figure 2.

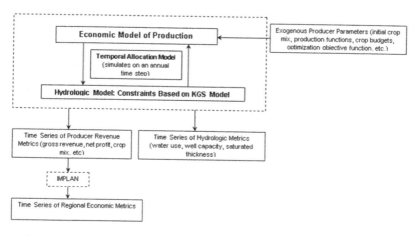

Figure 2.   (Color online) Flowchart of modeling process.

Temporal allocation model

The temporal allocation model is a dynamic optimization model that combines an economic model of producer choice (which predicts land, crop, and water allocations) with a hydrological model (which predicts well capacity and declines in saturated thickness), and investigates dynamic interactions between the two. The model generates a 61-year time-series representation of groundwater use, aquifer levels, irrigated acreage, and measures of economic productivity. This time-frame was chosen to match previous modeling results from the KGS.

Dynamic optimization models are often used to evaluate the temporal impacts of policy options. Gisser and Mercado (1973) were among the first to integrate economic theory and hydrological theory of groundwater flow into a single model. They conceptualized a single-cell aquifer, defined appropriate equations of motion, and provided a theoretical basis for evaluating the market solution. Models comparable to Gisser and Mercado (1973) have since been developed and applied to groundwater policy management scenarios by Gisser and Sanchez (1980), Gisser (1983), Feinerman and Knapp (1983), Golden, Peterson, and O'Brien (2008), and Amosson, Almas, Golden, et al. (2009).

In our research, KGS's hydrological model replaces the more conventional single-cell aquifer model. A representative producer then maximizes profit by choosing optimal allocation of land, crops, and water over a 61-year timeframe. As saturated thickness declines, depth to water increases, which increases pumping costs and diminishes well capacity. When production costs increase and water availability decreases, the producer can shift their cropping patterns to re-optimize profit under these new conditions. This profit maximization problem (similar to Amosson, Almas, Golden, et al. 2009) is constrained by assumptions that the GMD#3 board of directors made about realistic changes in cropping patterns.

This research not only considers both normal and drought weather scenarios, but also allows annual rainfall to vary across the 61 years within a given weather scenario. Figure 3 illustrates annual precipitation for the normal and drought weather scenarios in Subarea 1. These weather scenarios were provided by KGS, so we could maintain consistency with their results. The normal weather scenario was based on historical data, whereas the drought weather scenario was based on 75% of normal weather.

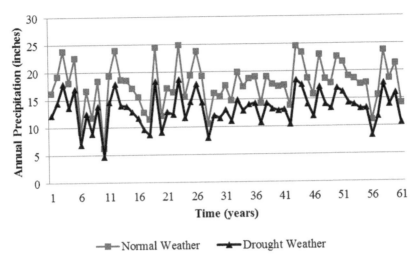

Figure 3. (Color online) Subarea 1 annual precipitation for Normal and Drought Weather scenarios. Source: Normal weather based on historical data from Kansas Geological Survey; drought weather is 75% of normal.

Variation in annual rainfall within a given weather scenario requires the temporal allocation model to contain dynamic crop production functions. These functions allow the relationship between inputs and outputs to change in response to annual changes in precipitation. The production function for irrigated corn, for example, would quantify the relationship between acre-inches of irrigation water applied and bushels of corn produced per acre. Stone, Schlegel, Khan, Klocke, and Aiken (2006) provides relationships between net irrigation and per acre crop yield, given various levels of annual precipitation, for the major irrigated crops in western Kansas. In each time period, these dynamic production functions, along with production costs and returns, precipitation, and current saturated thickness, are used to determine the profit-maximizing crop acres, crop yield, and groundwater consumption for that time period.

Regional economic impact model

When agricultural groundwater use is restricted, either from policy intervention or declining well capacity, crop production will likely be reduced in the near term and producers and local communities will thus incur negative economic impacts. These direct economic impacts will ripple through the economy, creating additional indirect and induced impacts (often called secondary impacts). The magnitude of these impacts will depend on the magnitude of groundwater-use reductions, current level of groundwater-use efficiency in the production process, number of acres affected, crop mix for those acres, shape of the crop-specific production functions (which are impacted by localized growing season characteristics such as precipitation and temperature), prices, costs, and the relative economic importance of agriculture in the affected communities. Following Amosson, Almas, Golden, et al. (2009), direct impacts for each policy-weather scenario are estimated by the temporal allocation models, which are then used as input to the regional economic impact model.

Impact Analysis for Planning (IMPLAN) software is used to conduct an input-output (I-O) analysis, which quantifies indirect and induced economic impacts to the regional economy. Given estimates of direct economic impacts, the IMPLAN software accounts for endogenous linkages between production, labor and capital income, trade, and household expenditures, and uses them to estimate impacts on sector output, value added, household income, and employment (Minnesota IMPLAN Group [MIG], 1999). Our two-model economic analysis captures not only the direct and indirect impacts on production, but induced impacts, as well. Direct impacts represent the initial effects of an outside shock on a particular sector. Indirect impacts represent changes in a particular sector's demands for intermediate goods. Induced impacts represent changes in demand for goods and services (i.e., spending) by households whose income has been altered.

IMPLAN analysis uses published government economic data to summarize financial transactions in a region at a specific point in time. The method generates 'multipliers' that reflect how industry sectors, households, and other institutions are financially linked one to another and to the overall economy, and how they are impacted by an exogenous economic shock. These multipliers can then be used to determine the size and direction of secondary economic impacts.

The literature suggests that IMPLAN industry production functions, based on national averages, may not be appropriate for regional analyses. MIG (1999) suggests that, because IMPLAN's agriculture data is derived rather than directly observed, researchers with better data should incorporate that data when building their IMPLAN models. Analysis by parts (ABP) is a means of incorporating local information by creating an industry production function that specifies first-round indirect impacts. We use ABP to estimate multipliers for each of the irrigated and non-irrigated crops. These weighted average multipliers (with weights based on crop mix) are then used to calculate first-round indirect impacts in each time period. First-round indirect impacts are then added to direct impacts to derive a total impact estimate.

This research estimates groundwater use, direct economic impacts to agriculture, and secondary economic impacts to the community over a 61-year timeframe. However, IMPLAN uses a single year's data to create the structural matrices, production functions, and multipliers that describe a regional economy. Unfortunately, crop prices and costs, which determine the magnitude of various multipliers, vary significantly across time. It is therefore important to select the appropriate annual IMPLAN dataset to ensure that the chosen year does not contain unwanted anomalies in prices and costs (Thorvaldson & Prichett, 2007). For example, if had we used IMPLAN data from the year 2000 to estimate economic impacts for the period 2001 through 2005, we would have overestimated impacts to Total Value Added by over 65%. To prevent a single year's anomalies from biasing our results, we use IMPLAN data for the years 2002 through 2009 to construct average multipliers.

IMPLAN generates several types of outputs that quantify the total economic impact of an external shock; all of these outputs are broken down into direct, indirect, and induced effects. This study reports Total Industry Output and Value Added. Total Industry Output is the total value of industry output for a given time frame (MIG, 1999); it can be loosely interpreted as the value of sales. Value Added consists of four components: (1) employment compensation (wage, salary, and benefits paid by employers), (2) proprietor income (payments received by self-employed individuals as income), (3) other property income (payments to individuals in the form of rents), and (4) indirect business taxes (basically all taxes with the exception of income tax).

Researchers often report employment impacts estimated by IMPLAN, even though Thorvaldson and Prichett (2007), Hughes, (2003), and Norvell and Kluge (2005) suggest that IMPLAN may over-estimate them. Golden et al. (2008) found that the most likely employment impacts were approximately 7.7% of the employment impacts initially generated by IMPLAN. Due to this ambiguity, we do not report employment impacts. Value Added includes employment compensation, however, which begs the question of whether IMPLAN might over-estimate this measure of impact as well. None of the research cited above speaks to this question, so additional research is needed in this area.

Net present value analysis

Net present value analysis is a standard method used to compare long-term projects. It discounts future cash flows to present values and sums the resulting income stream. Net present value is a reasonable method for long-lived entities to use when comparing investments or project costs. However, Ferejohn and Page (1978) argue that the use of discounted present value is inappropriate when dealing with welfare maximization over an infinite horizon because it implies that underlying social preferences remain constant over time. Gisser (1983) suggests that net present value theory is only appropriate under the assumption that the present generation feels altruistic toward future generations and therefore represents their best interest.

Net present value calculations require a 'discount rate' that transforms future values into present values. The use of a positive discount rate implies the conventional view that profits today are more valuable than profits in the future. A positive discount rate might be chosen by a producer who focuses on the near-term cash flows necessary to meet current obligations, such as land and equipment payments. A 0% discount rate implies neutrality about the timing of cash flows. A negative discount rate implies that profits, and by extension water, is valued more highly in the future than it is today. Such a stance might be taken by a producer who wants to conserve water resources today so their children can enjoy the stability of irrigated production in the future. We assume a 0% discount rate to compare policy alternatives, but we also conduct a sensitivity analysis to see how alternative discount rates affect our results.

Groundwater valuation

The purpose of this research is to compare policy options for reducing groundwater consumption. We expect a policy such as the GMD#3 Reallocation scenario to restrict water use relative to the Status Quo scenario, and result in less total groundwater consumed over a 61-year timeframe. Few temporal allocation studies estimate the value of the conserved groundwater (Amosson, Almas, Golden, et al., 2009; Golden et al., 2008). This may be because, from a purely agricultural production standpoint, groundwater has no value until it is brought to the surface and used to produce agricultural products. Additionally, studies that discount future values by assuming a positive discount rate may find that that any remaining water in the future (after 61 years) has negligible value today. Amosson, Almas, Bretz, et al. (2006) suggests that the cost of generating water savings must be weighed against the benefit of doing so. To accomplish this, however, a 'price tag' needs to be given to conserved water.

It is relatively straight-forward to compare policy scenarios based on differences in producer revenues, well-capacity, and saturated thickness. It is less obvious how to value differences in total groundwater consumed. One method is to measure differences in monetary returns between irrigated and dryland agricultural production, divided by the quantity of water used in irrigation (Leatherman, Golden, Featherstone, Kastens, & Dhuyvetter, 2006; Pritchett, Watson, Thorvaldson, & Ellingson, 2005; Supalla, Buell, & McMullen, 2006). In accordance with this approach, we measure the average value of applied groundwater (from a producer's perspective) as the difference in undiscounted cumulative net returns from the scenario that uses groundwater (Scenario 1 or 3) versus the Immediate Conversion to Dryland scenario (Scenario 2), divided by the cumulative groundwater used in the former scenario. We measure the average value of applied groundwater (from the regional economy's perspective) as the difference in undiscounted cumulative Total Industry Output or Value Added from the scenario that uses groundwater (Scenario 1 or 3) versus the Immediate Conversion to Dryland scenario (Scenario 2), divided by cumulative groundwater used in the former scenario. This average value per acre-foot of groundwater used is then assumed to be the value of each acre-foot of groundwater conserved.

Irrigated crop productivity growth

Throughout history, adoption of new crop varieties, cultural practices, and biotechnology has allowed producers to increase yields and decrease input use. These improvements have impacted both irrigated and non-irrigated crops. Most temporal allocation studies assume that growth rates for irrigated and non-irrigated crop yields are equal. However, a review of historical National Agricultural Statistics Service data suggests that yields for many irrigated crop may be increasing faster than for non-irrigated crops.

Irrigated crop producers using these advances can increase yield per acre while holding water use constant. As yield differentials between irrigated and non-irrigated crops increase, *ceteris paribus*, the value of groundwater increases. Figure 4 illustrates how the value of groundwater, as measured by gross returns to corn, has increased across time. These data reflect the difference in average irrigated corn yield and average non-irrigated corn yield as reported by the National Agricultural Statistics Service for Crop Reporting District #30 (southwest Kansas) multiplied by prices reported by the Economic Research Service. Between 1975 and 2011, this value differential increased at an average rate of 3.5% per year. Between 1975 and 1999, however, it increased at an average rate of 0.4%

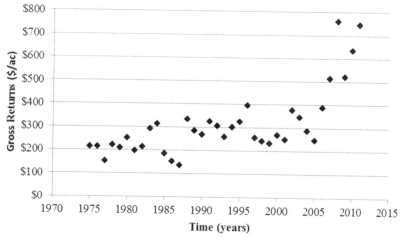

Figure 4. Historical values of groundwater as measured by the difference in gross returns for irrigated and non-irrigated corn.

Source: Data reflect the difference in average irrigated corn yield and average non-irrigated corn yield, as reported by National Agricultural Statistics Service for Crop Reporting District #30 (southwest Kansas), multiplied by prices reported by USDA-Economic Research Service.

per year, while between 2000 and 2011 it increased at an average rate of 9.7% per year. Little economic research has quantified the growth-rate differential between irrigated and non-irrigated crop yield. Amosson, Almas, Golden, et al. (2009) assumes all irrigated crop yields increase by 0.5% per year. We conservatively assume that irrigated crop yields increase by 0.5% per year relative to non-irrigated crop yields.

Data

Hydrology model

The KGS groundwater model includes 15,000 cells, each a square-mile in size; groundwater flow occurs in 12,083 of these cells. For each subarea, average results across cells in the KGS model were used as hydrological inputs to our temporal allocation model. The KGS model specifically provides annual water use, saturated thickness, depth to water, and percentage of land that becomes dry after extended pumping. These hydrological data were used to initially parameterize the temporal allocation model, as reported in Table 1.

Several economic studies of groundwater decline have demonstrated the impact of saturated thickness on well-capacity, and subsequently on crop choice, production costs and revenues, and the decision to cease irrigated crop production (Brill & Burness, 1994; Burness & Brill, 2001; Sloggett & Mapp, 1984). To place appropriate constraints on total groundwater availability, each of these studies has defined well-pumping capacity as a function of saturated thickness. We rely on the KGS model's assumed relationship between practical well-yield and saturated thickness. As saturated thickness diminishes, well-capacity declines and, in the long-run, reduces total water availability.

Temporal allocation model

The temporal allocation model requires crop production budgets. Prices and input costs for irrigated and non-irrigated crop production are based on five-year averages from the Kansas State University Annual Farm Management Guide, with slight modifications as recommended by GMD#3. Because yields vary with precipitation and irrigation water applied, and revenues and variable expenses vary with yield, net returns consequently vary from year to year. This variation, if different between policy-weather scenarios, may have implications for the regional economy.

Table 1. Hydrological parameters for priority subareas of southwest Kansas.

Item	High Priority Area		
	1	2	3
Recharge (inches/year)	1.94	2.24	1.01
Depth to Water (feet)	227.93	145.22	268.58
Saturated Thickness (feet)	222.55	120.43	227.37
Hydraulic Conductivity (ft/day)	43.64	53.06	46.29
Specific Yield	0.17	0.17	0.18
Average Well Capacity (gallons per minute)	633.19	489.74	665.31
Average Decline in Saturated Thickness (feet)	2.98	2.44	2.28
Average Water Use per Acre (feet)	1.45	1.32	1.13
Average Annual Water Use (acre feet)	178,284.6	115,994.6	145,964.0

Table 2.   IMPLAN multipliers for irrigated and dryland crops in southwest Kansas.

Category	Impact Type			
	Direct	Indirect	Induced	Total
Irrigated crops				
Total Industry Output	1.00	0.21	0.18	1.39
Value Added	0.37	0.13	0.11	0.61
Dryland crops				
Total Industry Output	1.00	0.18	0.25	1.42
Value Added	0.50	0.11	0.15	0.75

Regional economic impact model

Kansas State University's Annual Farm Management Guides, for the appropriate year, were used in the ABP-IMPLAN analysis to inform our development of multipliers. Examples of resulting multipliers for irrigated and non-irrigated crops are reported in Table 2. The total impact multiplier for non-irrigated crops is larger than that for irrigated crops. This is because a larger percentage of inputs for non-irrigated crops are obtained locally. This does not imply that an acre of non-irrigated crop production has a greater value to the community than an acre of irrigated crop. Revenue generated by an acre of irrigated crop is much larger than for an acre of non-irrigated crop, so the irrigated crop's associated impacts will also be larger.

Results

In groundwater policy analysis, a status-quo scenario is typically constructed that represents a baseline and assumes unconstrained producer behavior. A second scenario is constructed that represents the exogenous impact of a policy option, which imposes a constraint on producer behavior. The two scenarios' results are then compared to assess the impact of the exogenous shock. For our case, the exogenous shock of primary interest is the GMD#3 Reallocation scenario (scenario 3), which essentially restricts groundwater use to a pre-defined annual quantity. The following results compare the Reallocation scenario to the Status Quo scenario (scenario 1) for the two weather scenarios (normal versus drought).

Figure 5 illustrates Subarea 1's annual groundwater use, for scenarios 1 and 3, under drought versus normal weather. Under normal weather, the Reallocation scenario uses 9.5% less groundwater than the Status Quo scenario. Under drought weather, the Reallocation scenario uses 14.8% less groundwater than the Status Quo scenario. The Reallocation scenario restricts groundwater use to a pre-defined annual quantity. This pre-defined quantity is less than current usage, so the Reallocation scenarios use the same quantity of water, regardless of annual precipitation. Groundwater use in the Status Quo scenario, however, varies with annual precipitation. Under drought weather, the Status Quo scenario uses more groundwater in the short-run, relative to under normal weather, to compensate for low precipitation. Because more groundwater is used in the short-run, saturated thickness declines faster and well-capacity diminishes faster. These two factors, combined, result in less groundwater use in the long-run. Short-run refers to the first portion of the 61-year model time frame, and long-run refers to the latter portion of the 61-year model

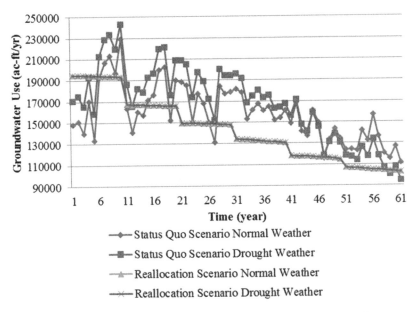

Figure 5. (Color online) Subarea 1 time-series for annual groundwater use (acre-feet per year).

Table 3. Cumulative groundwater use (million acre-feet) for Subarea 1 under normal versus drought weather, and under various policies.

Scenario	Cumulative Groundwater Use	Policy-induced Change in Groundwater Use[a]	Drought-induced Change in Groundwater Use[b]
Normal weather			
Status Quo (scenario 1)	9.67	0	NA
Immediate Conversion to Dryland (scenario 2)	0	−9.67	NA
Reallocation (scenario 3)	8.76	−0.91	NA
Drought weather			
Status Quo (scenario 1)	10.28	0	+0.61
Immediate Conversion to Dryland (scenario 2)	0	−10.28	0
Reallocation (scenario 3)	8.76	−1.52	0

Note: [a] e.g., (Normal weather, Scenario j) − (Normal weather, Scenario 1), where j = 1, 2, 3.
[b] e.g., (Drought weather, Scenario i) − (Normal weather, Scenario i), where i = 1, 2, 3.

time frame. Table 3 reports cumulative groundwater use for Subarea 1 under normal versus drought weather.

As saturated thickness diminishes, and well-capacity diminishes, pumping costs increase, and the producer has to pump more hours to maintain water use. At some point in time, the producer re-optimizes profit by adjusting their crop mix and irrigation activities. They may, for example, begin splitting limited irrigation water between two summer crops, or between a summer and winter crop (typically winter wheat). As well-capacity continues to decline, the producer will eventually cease irrigation and convert to dryland production. Irrigated winter wheat and non-irrigated crops are much less profitable than irrigated corn and alfalfa, however, and thus cause net revenue to decline as they are introduced into the crop mix. The producer's choice to shift acreage from one crop to another is endogenously determined by the temporal allocation model.

For Subarea 1, at the end of 61 years, the Status Quo scenario under normal weather has converted 62.7% of the initially-irrigated acreage to dryland crops and irrigated wheat. The Reallocation scenario, in contrast, has converted 50.0% of the initially-irrigated acreage to dryland crop and irrigated wheat. From this we can infer that, in general, the Reallocation scenario will maintain a more profitable crop mix in the long-run.

Figure 6 illustrates producer net revenue for the four scenarios that use irrigation water in Subarea 1 (i.e., Reallocation and Status Quo scenarios under normal versus drought weather). In the long-run, producer net revenue is indeed higher for the Reallocation scenario than the Status Quo scenario even though less groundwater is used. This is because the Reallocation scenario preserves well-capacity in the long-run, which allows for a more profitable crop mix. The Reallocation scenario increases cumulative net revenue (after valuing conserved groundwater) by 4.3% under normal weather, and by 12.8% under drought, relative to the Status Quo.

Under normal weather, the average value of cumulative groundwater used (from the producer's perspective) is $166.55 per acre-foot in the Status Quo scenario, versus $174.78 per acre-foot in the Reallocation scenario (Table 4). The average value of groundwater used is higher for the Reallocation scenario because less groundwater is used in the production of higher valued crops (i.e., in the calculation of average value, the denominator is smaller and the numerator is larger). Under drought weather, the average value of cumulative groundwater used is $161.38 per acre-foot in the Status Quo scenario, versus $182.17 per acre-foot in the Reallocation scenario (Table 4). The value of cumulative groundwater used is again higher for the Reallocation scenario because less groundwater is used in the production of higher valued crops.

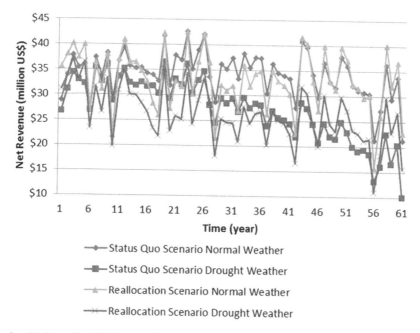

Figure 6. (Color online) Subarea 1 time-series for producer net revenue (US$).

Table 4. Producer's cumulative net revenue (in million US$) for Subarea 1 under normal versus drought weather, and under various policies.

Scenario	Cumulative Net Revenue	Policy-induced Change in Net Revenue[a/]	Average Value of Water Used[b/]	Value of Conserved Water	Net Producer Impact
Normal weather					
Status Quo (scenario 1)	$2093.1	$0	$165.55	$0	$0
Immediate Conversion to Dryland (scenario 2)	$491.1	−$1602.0	NA	NA	NA
Reallocation (scenario 3)	$2021.4	−$71.7	$174.78	$160.9	$89.3
Drought weather					
Status Quo (scenario 1)	$1664.9	$0	$161.38	$0	$0
Immediate Conversion to Dryland (scenario 2)	$6.3	−$1658.6	NA	NA	NA
Reallocation (scenario 3)	$1601.3	−$63.6	$182.17	$277.2	$213.6

Note: [a/] e.g., (Normal weather, Scenario j) – (Normal weather, Scenario 1), where j = 1, 2, 3.
[b/] Expressed in dollars per acre-foot, rather than million US$.

Table 5. Cumulative total industry output (million US$) for Subarea 1 under normal versus drought weather, and under various policies.

Scenario	Cumulative Total Industry Output	Policy-induced Change in Total Output[a/]	Average Value of Water Used[b/]	Value of Conserved Water	Net Industry Impact
Normal weather					
Status Quo (scenario 1)	$10,307.7	$0	$857.01	$0	$0
Immediate Conversion to Dryland (scenario 2)	$2,015.0	−$8,292.7	NA	NA	NA
Reallocation (scenario 3)	$10,077.9	−$229.9	$920.88	$847.9	$618.0
Drought weather					
Status Quo (scenario 1)	$8,931.1	$0	$760.00	$0	$0
Immediate Conversion to Dryland (scenario 2)	$1,120.3	−$7,810.8	NA	NA	NA
Reallocation (scenario 3)	$8,917.7	−$13.4	$890.56	$1,355.2	$1,341.7

Note: [a/] e.g., (Normal weather, Scenario j) – (Normal weather, Scenario 1), where j = 1, 2, 3.
[b/] Expressed in dollars per acre-foot, rather than million US$.

Total Industry Output is a measure of gross revenues for both the producer and industry sectors supplying the producer. While the producer's economic well-being is largely determined by net revenues, the economic well-being of rural communities is sensitive to Value Added which is determined by Total Industry Output. Table 5 reports cumulative Total Industry Output for Subarea 1 under the two weather scenarios, after valuing conserved groundwater. Compared to the Status Quo scenario, the Reallocation scenario increases cumulative Total Industry Output (plus the value of conserved water) by 6.0% under normal weather, and by 15.0% under drought. Water conservation therefore has a more positive impact on Total Industry Output during a drought. This is because the Reallocation scenario maintains a higher-valued irrigated crop mix in the long-run, compared to Status Quo, particularly during drought. Thus, gross revenue from irrigated production and input purchases are not as heavily impacted by drought in the Reallocation scenario.

The average value of cumulative groundwater used (from the community's perspective, as measured by Total Industry Output), under normal weather, is $857.01 per acre-foot for the Status Quo scenario versus $920.88 per acre-foot for the Reallocation scenario (Table 5). The average value of groundwater used is higher (from the community's perspective) for the Reallocation scenario because less groundwater is used in the production of higher valued crops. Under drought, the average value of groundwater used is $760.00 per acre-foot for the Status Quo scenario, versus $890.56 per acre-foot for the Reallocation scenario. The implication is that water conservation (i.e., Reallocation) has a larger positive impact on cumulative Total Industry Output during drought than during normal weather.

Value Added is closely related to the combined net revenues for the producer and the industry sectors supplying them. The economic well-being of rural communities responds to Value Added generated within those communities. Table 6 reports cumulative Value Added for Subarea 1 under normal versus drought weather, after valuing conserved groundwater. The Reallocation scenario increases Value Added by 5.4% under normal weather, and by 8.6% under drought. Again, water conservation policy has a more positive impact on Value Added during a drought.

Under normal weather, the average value of groundwater used (from the community's perspective, as measured by Value Added) is $310.63 per acre-foot for the Status Quo scenario, and $355.60 per acre-foot for the Reallocation scenario (Table 6). Under drought, the average value of groundwater used is $300.27 per acre-foot for the Status Quo scenario, and $353.78 per acre-foot for the Reallocation scenario (Table 6).

Table 7 summarizes, for all three subareas (as opposed to just Subarea 1), differences in cumulative groundwater use, cumulative producer net revenue, Total Industry Output, and Value Added between Status Quo and Reallocation scenarios. Two general patterns emerge. First, changes induced in the Reallocation scenario (relative to the Status Quo scenario) are generally larger (in absolute magnitude) under drought than normal weather. This reflects that, as water becomes increasingly scarce, the potential gains or losses from water management policy are magnified. Second, the Reallocation scenario's impacts vary notably across Subareas 1, 2 and 3. For example, average value of conserved groundwater to the producer (as measured by net revenue; though not included in Table 7) ranges from $129.03 to

Table 6. Cumulative value added (million US$) for Subarea 1 under normal versus drought weather, and under various policies.

Scenario	Cumulative Total Industry Output	Policy-induced Change in Value Added[a/]	Average Value of Water Used[b/]	Value of Remaining Water	Net Industry Impact
Normal weather					
Status Quo (scenario 1)	$4517.0	$0	$310.63	$0	$0
Immediate Conversion to Dryland (scenario 2)	$1511.3	−$3005.8	NA	NA	NA
Reallocation (scenario 3)	$4449.7	−$67.4	$335.60	$309.0	$241.6
Drought weather					
Status Quo (scenario 1)	$3926.2	$0	$300.27	$0	$0
Immediate Conversion to Dryland (scenario 2)	$840.2	−$3086.0	NA	NA	NA
Reallocation (scenario 3)	$3937.8	$11.6	$353.78	$325.7	$337.3

Note: [a/] e.g., (Normal weather, Scenario j) − (Normal weather, Scenario 1), where j = 1, 2, 3.
[b/] Expressed in dollars per acre-foot, rather than million US$.

Table 7.   Impacts of the Reallocation scenario relative to the Status Quo scenario, for each subarea, under normal and drought weather.

Metric	Normal Weather	Drought Weather
Cumulative Groundwater Use		
Subarea 1	−9.5%	−14.8%
Subarea 2	−31.7%	−31.6%
Subarea 3	−14.7%	−19.6%
Cumulative Net Producer Revenue		
Subarea 1	4.3%	12.8%
Subarea 2	−0.9%	5.7%
Subarea 3	−1.3%	1.6%
Cumulative Total Industry Output		
Subarea 1	6.0%	15.0%
Subarea 2	5.8%	11.1%
Subarea 3	0.8%	3.6%
Cumulative Value Added		
Subarea 1	5.3%	8.6%
Subarea 2	−1.2%	3.6%
Subarea 3	−2.0%	−2.8%

$182.17 per acre-foot by subarea. Average value of conserved groundwater to the rural community (as measured by Total Industry Output) ranges from $644.56 to $920.88 per acre-foot by subarea. Variation in subarea-specific results are due to differences in initial hydrological conditions, the magnitude of water-use restrictions, and dryland production options, which determine how the irrigated crop-mix varies over time and the rate at which irrigated cropland is converted to dryland production.

Table 7 also shows that, from a producer's perspective (as measured by cumulative producer net revenue), the Reallocation scenario performs better under drought weather than the Status Quo in all subareas. It also performs better under normal weather in Subarea 1, but not Subareas 2 or 3. In general, positive impacts of the Reallocation scenario can be attributed to restriction of groundwater use in the short-run, which maintains well-capacity in the long-run, and thus allows producers to maintain a more profitable crop-mix in the long-run. As mentioned before, benefits to the producer of groundwater conservation are enhanced in times of drought. The same conclusions hold true from the communities' perspective, as measured by Total Industry Output.

While these results are sensitive to assumptions regarding the future value of groundwater (i.e., the discount rate), they suggest that groundwater-use restrictions imposed by the Reallocation scenario may lead to economic benefits for both the producer and rural economies, depending on their subarea. These results also suggest that the economic impacts of future drought conditions on producers and rural economies may be mitigated in some subareas if the Reallocation scenario is implemented today.

'Sensitivity analysis' is a method for testing the robustness of a model's output to different model inputs. Golden et al. (2008) used sensitivity analysis to investigate the sensitivity of model results to alternative discount rates. The Reallocation scenario's impacts under normal weather, relative to the Status Quo scenario, vary depending on the assumed discount rate (Table 8). Results for Subareas 1 and 3 are relatively insensitive to the discount rate, averaging roughly a 0.5% change per 1% change in discount rate. Results for Subarea 2 are very sensitive to the discount rate, averaging approximately a 12.5% change per 1% change in discount rate. The sensitivity of Subarea 2 can be linked to two

Table 8. Impacts of the Reallocation scenario, relative to the Status Quo scenario, for each subarea under normal weather, assuming alternative discount rates.

Metric	Discount Rate				
	3%	1.5%	0%	−1.5%	−3%
Cumulative Net Producer Revenue					
Subarea 1	2.3%	3.2%	4.3%	5.3%	6.4%
Subarea 2	−30.2%	−14.8%	−0.9%	10.8%	20.0%
Subarea 3	−3.3%	−2.2%	−1.3%	−0.6%	0.1%
Cumulative Total Industry Output					
Subarea 1	4.2%	5.1%	6.0%	6.8%	7.6%
Subarea 2	−18.0%	−5.6%	5.8%	15.6%	23.5%
Subarea 3	−1.9%	−0.4%	0.8%	1.8%	2.6%
Cumulative Value Added					
Subarea 1	1.4%	3.4%	5.3%	7.0%	8.5%
Subarea 2	−34.5%	−16.9%	−1.2%	11.8%	21.9%
Subarea 3	−9.1%	−5.1%	−2.0%	0.4%	2.2%

unique characteristics: (1) the magnitude of water-use restriction imposed, which generates less groundwater use and less profit in both the short-run and long-run, regardless of the weather scenario, and (2) over 56% of the subarea's acreage has rather poor soils, not suitable for dryland production.

Conclusions

The purpose of this research is to provide input into the water planning process for high priority subareas in southwest Kansas. The study considers three groundwater-use scenarios (Status Quo, Immediate Conversion to Dryland, and Reallocation) and two weather scenarios (normal and drought). Stakeholder input suggests that a reduction in groundwater use may be desirable to preserve the Ogallala Aquifer and extend its economic contribution to both producers and the regional economy. This research estimates measures of cumulative producer net revenue, cumulative regional Total Industry Output, and cumulative regional Value Added in an attempt to identify the 'best' water conservation policy. Additionally, this research estimates the monetary value of conserved groundwater, and attempts to account for future growth in irrigated crop productivity.

We integrate information from a hydrological model into two economic models (temporal allocation, and regional economic impacts), which were developed to estimate impacts over a 61-year time horizon. Economic models that attempt to predict the future are, by their very nature, subject to error; results are most appropriately viewed as a 'best guess'. Our impact estimates are based on a variety of assumptions; a different set of assumptions will alter the magnitude of impacts. So long as consistent assumptions are maintained across policy options, and stakeholders are comfortable with those assumptions, small changes in assumptions might not alter the relative order of policy impacts.

Our results are sensitive to assumptions regarding the future value of groundwater and discount rates, but generally suggest that groundwater-use restrictions imposed by the Reallocation scenario may economically benefit both producers and rural economies in some subareas. Results suggest, however, that severe water-use restrictions, such as those modeled for Subarea 2 may lead to significant economic losses. This particular Reallocation scenario imposed the same desired future aquifer conditions on all subareas, but more severe restrictions were necessary in Subarea 2 to achieve them. Our results for Subarea 2 suggest that a one-size-fits-all policy prescription may not be appropriate. Variation in subarea-specific results are due to differences in the magnitude of water-use restrictions, initial hydrological conditions, and dryland production options, which determine how the irrigated crop-mix varies over time and the rate at which irrigated cropland is converted to dryland production. Results generally suggest though that, if future drought conditions exist, economic impacts on producers and rural economies may be mitigated if the Reallocation scenario is implemented today.

Adoption of the Reallocation scenario's water conservation policy may reduce groundwater consumption in the short-run, but will not reduce groundwater consumption over an infinite horizon. Even with severe reductions in groundwater use today, subareas of southwest Kansas will remain over-appropriated, and water saved today will eventually be used, thus exhausting the water resource.

Caveats and limitations

Cumulative estimates of Total Industry Output and Value Added are subject to a degree of uncertainty. IMPLAN is normally considered a short-run analytical tool because it makes an implicit assumption that the structure of the rural economy stays constant over the timeframe analyzed. Although we have stated clearly the assumptions on which our impacts are calculated, and have reported the calculated impacts accurately, their exact magnitude may be in question. However, all scenarios are based on the same assumptions, so our policy comparison is appropriate.

This study estimates the profit impact of groundwater conservation policy under varying climatic conditions, but does not address the potential for increased risk exposure. Local producers have indicated that increased risk exposure is their primary concern when considering the adoption of limited irrigation in response to groundwater conservation policy. Additional research is needed to identify and quantify the risks associated with limited irrigation in our study area.

Lastly, when groundwater conservation policies are implemented, less groundwater is typically used than in the status quo. This research places an average monetary value on conserved groundwater. These values are based on the assumption that prices, yields, cultural practices, and technology remain constant over time, and that the producer and communities' discount rate is 0%. If significant gains are made in water-use productivities, in excess of the assumed 0.5% annual growth, the monetary values presented in this research may be significantly understated. This research illustrates the importance of placing an economic value on conserved groundwater when conducting policy analysis. Inclusion of conserved water's value changes both the magnitude and sign of projected policy impacts, and changes the policies' relative performance. Exclusion of conserved water's value could lead to erroneous policy choices.

Acknowledgements

This research was funded by the State of Kansas Water Plan Fund, Groundwater Management District #3, and the USDA-ARS Ogallala Aquifer Project.

References

Amosson, S. H., Almas, L. K., Bretz, F., Gaskins, D., Guerrero, B., Jones, D., Marek, T., New, L., & Simpson, N. (2006). *Water management ttrategies for reducing irrigation demands in Region A*. Texas Water Development Board. Retrieved from http://www.twdb.state.tx.us/home/index.asp

Amosson, S., Almas, L., Golden, B., Guerrero, B., Johnson, J., Taylor, R., & Wheeler-Cook, E. (2009, June). *Economic impacts of selected water conservation policies in the Ogallala Aquifer* (Staff Paper No. 09-04). Kansas State University Agricultural Experiment Station and Cooperative Extension Service.

Brill, T. C., & Burness, H. S. (1994). Planning versus competitive rates of groundwater pumping. *Water Resources Research, 30*(6), 1873–1880.

Burness, H. S., & Brill, T. C. (2001). The role of policy in common pool groundwater use. *Resource and Energy Economics, 23*, 19–40.

Feinerman, E., & Knapp, K. C. (1983). Benefits from groundwater management: Magnitude, sensitivity, and distribution. *American Journal of Agricultural Economics, 65*, 703–710.

Ferejohn, J., & Page, T. (1978). On the foundations of intertemporal choice. *American Journal of Agricultural Economics, 60*(2), 269–275.

Gisser, M. (1983). Groundwater: Focusing on the real issue. *Journal of Political Economy, 91*(6), 1001–1027.

Gisser, M., & Mercado, A. (1973). Economic aspects of ground water resources and replacement flows in semiarid agricultural areas. *American Journal of Agricultural Economics, 55*(3), 461–466.

Gisser, M., & Sanchez, D. A. (1980). Competition versus optimal control in groundwater pumping. *Water Resources Research, 16*(4), 638–642.

Golden, B., Peterson, J., & O'Brien, D. (2008, February). *Potential economic impact of water use changes in Northwest Kansas* (Staff Paper No. 08-02). Kansas State University Agricultural Experiment Station and Cooperative Extension Service.

Hughes, D.W. (2003). Policy use of economic multiplier and impact analysis. *Choices, 18*(2), 25–29.

Leatherman J., Golden, B., Featherstone, A., Kastens, T., & Dhuyvetter, K. (2006, May). *Regional economic impacts of the conservation reserve enhancement program in the Upper Arkansas River Basin.* Final report submitted to the Kansas Water Office, reported to the House committee on Natural Resources. Retrieved from http://www.agmanager.info/policy/water/13-Leatherman & Golden-CREP.PDF

Minnesota IMPLAN Group, Inc. (1999). *IMPLAN Professional Software, Analysis, and, Data Guide.* Stillwater, MN: Author. Retrieved from http://www.implan.com

Norvell, S., & Kluge, K. (2005, July). *Socioeconomic impacts of unmet water needs in the Panhandle Water Planning Area,* Texas Water Development Board. Retrieved from http://www.twdb.state.tx.us/home/index.asp

Pritchett, J., Watson, P., Thorvaldson, J., & Ellingson, L. (2005). Economic impacts of reduced irrigated acres: Example from the Republican River Basin. *Colorado Water, February,* 4–6. Retrieved from http://www.cwi.colostate.edu/newsletters/2005/ColoradoWater_22_1.pdf

Sloggett, G. R., & Mapp, H. P. (1984). An analysis of rising irrigation costs in the Great Plains. *Water Resources Bulletin, 20*(2), 229–233.

Stone, L. R., Schlegel, A. J., Khan, A. H., Klocke, N. L., & Aiken, R. M. (2006). Water supply: Yield relationships developed for study of water management. *Journal of Natural Resources & Life Sciences Education, 35*, 161–173.

Supalla, R., Buell, T., & McMullen, B. (2006). *Economic and state budget cost of reducing the consumptive use of irrigation water in the Platte and Republican Basins.* Unpublished research report. Lincoln: University of Nebraska, Department of Agricultural Economics for the Nebraska Department of Natural Resources. Retrieved from http://digitalcommons.unl.edu/cgi/viewcontent.cgi?article=1042&context=ageconworkpap

Thorvaldson, J., & Prichett, J. (2007, July). *Economic impact analysis of reduced irrigated acreage in Four River Basins in Colorado.* Colorado Water Resource Research Institute. Retrieved from http://limitedirrigation.agsci.colostate.edu/present/economic_activity.pdf

Community adaptation to climate change: exploring drought and poverty traps in Gituamba location, Kenya

Amy Sherwood

Globally, many communities are vulnerable to weather-pattern variability. Climate change will act as a threat multiplier by increasing this variability. To combat growing vulnerability, strategies for adaptation must be developed. This study uses interviews and participatory research techniques to examine the effects of a year-long drought on women and poverty dynamics in Gituamba location, Kenya. It concludes that drought has the ability to create poverty traps and produce a poverty of time and energy among women. Some possible adaptation strategies include livelihood diversification, creation of cooperatives, conservation farming, and rehabilitation of communal boreholes.

Introduction

Climate researchers predict that variability and uncertainty in weather patterns will increase in coming years if anthropogenic climate change is left unmitigated. Among other things, their models suggest that precipitation patterns will change, resulting in less-predictable rainfall, more-frequent severe weather events, and a greater risk of drought in many areas (Intergovernmental Panel on Climate Change [IPCC], 2007; Kumssa & Jones, 2010; Solomon, Qin, Manning, Marquis, Averyt, Tignor, Miller, & Chen 2007). This predicted increase in weather variability is particularly challenging in Africa where 70% of the population depends on agriculture as a source of livelihood, and over 95% of this agriculture is rain-fed (African Partnership Forum, 2007). Furthermore, most agrarian African populations lack access to resources necessary for safeguarding their assets, which exacerbates their vulnerability to erratic weather (Eriksen & O'Brien, 2007).

There is an urgent need to identify ways in which African communities can adapt to increased variability in weather patterns (IPCC, 2007; Liverman, 2011; Speranza, Kiteme, Ambenje, Wiesmann, & Makali, 2010). Prior research shows that adaptation measures must be tailored to address specific vulnerabilities of the community they are meant to safeguard. Existing sources of vulnerability determine a community's susceptibility to extreme weather events, such as drought. Understanding this vulnerability, and reducing it through adaptation, is just as important as understanding how weather patterns will change (Liverman, 2011). Vulnerability to risk is increasingly accepted as a central characteristic of poverty. As such, development policies should be cognizant of, and informed by, existing sources of vulnerability and mechanisms that communities employ to mitigate it (Bhattamishra & Barrett, 2010).

Vulnerability is correlated not only with exogenous factors, such as resource availability, but also endogenous societal factors (Adger et al., 2007). The appropriateness of adaptation strategies therefore depends on the diverse options, values and goals within communities (Adger, Dessai, et al., 2007). Yohe and Tol (2002) identify eight community-specific criteria that influence adaptive capacity, including available technology, natural resources, structure of local institutions, human capital, social capital, processes employed by local decision-makers, a community's perception of the source of stress and its severity, and their ability to spread risk. Collectively, these studies suggest that creating effective adaptation policy requires an understanding of vulnerability at a household and community level. One method to investigate such vulnerability is to examine households' coping mechanisms during times of extreme weather events. This study uses interviews and participatory research techniques to examine how women in Gituamba location, Kenya adjusted their activities to cope with a year-long drought that began in 2008.

Drought coping mechanisms and vulnerability

Vulnerability to drought could act as a major determinant of future poverty dynamics, particularly as global climate change continues to alter weather patterns. To cope with drought, households currently employ diverse and dynamic coping strategies (Smucker & Wisner, 2008). A study of the 1984 drought in Kenya found that households relied on food stores, switched to horticulture, sought off-farm employment, relied on relatives that lived in urban areas, and sold livestock and other assets (Kamau, Anyango, Gitahi, Wainaina, & Downing, 1989).

A household's coping mechanisms depend on characteristics such as gender and number of working-age household members (Chambers, 1989). A household with two or more people of working age may be less constrained than a household with just one worker because two sources of income could potentially be available. If work is scarce, one person can continue to carry out household chores, while the others pursue wage labor or other income-generating activities. Gender of household members could affect coping in numerous ways; for example, some forms of wage labor are not socially acceptable livelihood strategies for women (Eriksen, Brown, & Mick, 2005).

Coping mechanisms can also reflect a household's priorities and options during a crisis. Immediate economic or food deficiencies are often the first priority, followed by maintenance of the means for long-term livelihood generation. Some coping mechanisms are thus geared toward minimizing short-term threats while others are oriented towards maintaining a long-term livelihood source. As resources dwindle, however, difficult decisions must be made about how remaining time, energy and finances are allocated. As a crisis worsens, more extreme and diverse coping methods are used. In an emergency situation, households will employ strategies that completely disregard long-term priorities, such as the sale of land by subsistence farmers (Eriksen et al., 2005).

Though coping strategies depend, in part, on household characteristics and priorities, they also depend on community dynamics. In some communities, for example, consumption of certain wild foods is so stigmatized that they will not be consumed unless a state of emergency exists, whereas in other communities the consumption of wild foods is a first line of defense to maintain food security (Smucker & Wisner, 2008). Marginalized groups, such as the poorest members of a community or women, may be excluded from collective coping or risk management strategies (Bhattamishra & Barrett, 2010). This leaves groups with the most pre-existing constraints in even more calamitous situations, which exacerbates fundamental inequalities. Societal values and hierarchies can therefore impede or facilitate adaptation and risk management strategies (Adams, Cekan, & Sauerborn, 1998; Bhattamishra & Barrett, 2010; Eriksen et al., 2005).

Poverty traps

People living in poverty are less able to make meaningful choices to improve their lives (Sen, 1999). A household living in extreme poverty must focus solely on survival and cannot afford to expand or diversify their livelihood options and income streams. Poverty trap theory asserts that households living in poverty lack access to assets necessary for lifting themselves out of poverty. Processes that exclude a household from these assets can differ by location and may include governmental oppression, geographic isolation, limited market access, disease, or erratic weather patterns (Sachs, 2005). As such, interventions or assets needed to lift a household or community out of a poverty trap can also vary greatly. It is imperative that poverty alleviation researchers understand and account for location or community-specific constraints and challenges (Barret, Marenya, McPeak, Minten, Murithi, Oluoch-Kosura, & Wangila, 2006).

A poverty trap dynamic also implies that, for households living in poverty, economic shocks such as drought can have long-term adverse effects on well-being. Impoverished households are implicitly underinsured against such risks. Shocks to their livelihoods therefore have the potential to weaken or reduce a household's labor force, diminish productive asset bases and deplete capital reserves. For an already impoverished household, these conditions can evolve into crises independent of the initial shock (Barrett et al., 2006; Little, Stone, Mogues, Castro, & Negatu, 2006). A household weakened due to insufficient nutrition, for example, may become more susceptible to chronic disease and increased long-term medical expenses. Such crises emerge as new structural conditions that hamper efforts at economic growth and reinforce the poverty dynamic. Poverty trap theory assumes a positive correlation between wealth and returns on assets. So, as the rich continue to grow in wealth (domestically or abroad), the economic position of people living in poverty traps holds constant or declines (Azaradis, 2006; Sachs, 2005).

Objectives

This paper explores relationships between poverty traps and household coping mechanisms through a case study that examines the effects of a year-long drought on women in Gituamba, Kenya. Using participatory research techniques, this study investigates the experiences of drought survivors and explores the following hypotheses:

(1) Drought has the potential to create poverty traps in communities that depend on rain-fed agriculture.
(2) Women may be disproportionately affected by drought.
(3) Drought coping mechanisms reveal community vulnerabilities and resiliencies, and can thus inform adaptation policy.
(4) Adaptation policy should go beyond relief operations and initiate strategies that reduce vulnerability to future erratic weather patterns.

Materials and Methods

Description of 2008–2009 drought

In 2008, the short rains that normally occur from October to December failed throughout Kenya. As it became apparent that the long rains of 2009 would also fail in many parts of the country, resulting in crop failure and high livestock death, a food security crisis began to develop for millions of Kenyans. Rapid assessment reports conducted by the Kenyan Red Cross indicated that as many as ten million Kenyans could be at risk of starvation. In mid-January 2009, the Kenyan government declared a state of emergency and began to appeal to the international donor community for food aid. In addition to food shortages, many areas of the country experienced extensive drying of water sources, inflated food prices, and human-animal conflicts over resources (Kenya Red Cross, 2009; World Food Program, 2009).

In Laikipia district of the Rift Valley province in Kenya, where Gituamba is located, the situation was similar: the 2008 short rains failed in most of the district, and the 2009 long rains faltered, causing a decrease in both quantity and duration of rainfall. This decrease in overall quantity, paired with the erratic nature of the remaining rainfall, led to widespread food insecurity and water shortages throughout Gituamba (Kaguara, Beethoven, Matere, & Koskei, 2009). In early 2010, Gituamba experienced dramatic improvements in food security. Although most households had still not experienced a major harvest since the 2008–2009 drought, conditions were improving, as people began to harvest and livestock became healthier. It was during this recovery phase that fieldwork for this study was conducted.

Participatory research methods

Participatory rural appraisal (PRA) describes a diverse set of methods that provide poor people with a means to categorize, analyze, and evaluate their lives. It is based on the notion that poor people have a fundamental right to conduct their own analysis of the challenges they face (Chambers, 1995). PRA methods stipulate that researchers should seek to learn from the poor, and avoid imposing their own ideas and opinions during this process. Specific goals of PRA include gaining direct knowledge from local people, decreasing biases, optimizing tradeoffs, and postulating potential solutions (Chambers, 1994).

Participatory Poverty Assessment (PPA) is a branch of PRA that uses participatory appraisal techniques to redefine poverty and priorities through the experiences of those living in it (Brock & McGee, 2002; Chambers, 1995). Ideally, PPA enables poor people to express their experiences, needs, and desires, and subsequently influence the design of policy that addresses root causes of poverty and vulnerability within their community (Chambers, 1994; Narayan, Chambers, Shah, & Petesch, 1999). PPA has gained credibility in recent years through its use in major development efforts, such as the World Bank's 'Voices of the Poor' project (Narayan, Patel, Schafft, Rademacher, & Koch-Schulte 1999).

This study used participatory research techniques to solicit in-depth personal accounts from survivors of the 2008–2009 drought. As primary investigator, I participated in daily livelihood and recovery strategies alongside the women of Gituamba. I engaged study participants in threat and priority-ranking exercises. I also developed interview questions and well-being indicators in conjunction with residents of Gituamba.

Semi-structured interviews

This study's data and conclusions are drawn from participant observation, informal focus-group discussions with women's self-help groups, and 40 semi-structured interviews.

Interviews involved both open-ended questions and a collection of indicators related to poverty and drought. Interviews consisted of four sections: (1) general demographic data, (2) poverty and life during normal years in Gituamba, (3) coping mechanisms during the 2008–2009 drought, and (4) drought's effects on women. A snowball sampling method was used, in which prior contacts were used to solicit initial participants, and then further participants were generated through each interviewee. This approach was chosen for its ability to seek out participants who would feel comfortable sharing in-depth accounts of their lives and the 2008–2009 drought. Interviews were conducted primarily in Swahili, which the primary investigator spoke only conversationally, so a local translator was used.

Results

Household dynamics in non-drought years

In non-drought years, the primary source of livelihood for 37 of the 40 respondents was farming maize and beans. Many respondents supplemented agricultural operations with at least two secondary livelihood strategies, which most commonly included livestock keeping, casual labor, small-scale sale of household necessities, or horticulture operations. Women faced many barriers to successful income generation even in non-drought years, such as a lack of capital to purchase proper agricultural inputs, a lack of labor to properly look after crops, low wages for casual labor, crop diseases, underdeveloped transportation infrastructure, and a lack of access to markets.

When ranking and discussing threats, respondents spoke most frequently and passionately about the lack of capital, which forced women to farm with inadequate fertilizer and non-certified seed, resulting in low yields. Even if a household were able to scrape together the capital needed for sufficient inputs, unreliable markets and unfair trade relationships meant there was no guarantee they would experience a return on their investment. This uncertainty surrounding returns on staple crop agriculture was the reason most women pursued the secondary strategies mentioned above. Secondary strategies not only supplemented income but also reduced household vulnerability to threats associated with staple crop agriculture.

Despite the lack of capital, respondents' priority rankings in non-drought years suggest they were striving to improve their household's situation. In non-drought years, the most common priorities were education of children (or grandchildren), acquiring more assets such as land and livestock, starting or investing in small businesses, and completing household improvements (e.g., switching from a thatched roof to a metal one). Only two of the respondents listed food-acquisition as a first priority in non-drought years.

The fact that most respondents were able to focus their resources on priorities that expanded livelihoods or improved quality of life signals that, in non-drought years, women of Gituamba were not caught in poverty traps. They were able to successfully allocate their time, capital and energy to afford improvements in their household's income-generating activities and well-being. For most respondents, their household poverty dynamic more closely followed classical economic theory, which suggests that impoverished people experience adequate returns on assets to continually improve their situation, if only marginally (Barret, Marenya, McPeak, Minten, Murithi, Oluoch-Kosura, & Wangila, 2006).

Drought and poverty traps

The 2008–2009 drought affected respondents and their families in diverse ways, but there were three fundamental ways in which it helped create potential poverty traps. First,

Table 1. Coping mechanisms used by women in Gituamba, Kenya during the 2008–2009 drought.

Coping Mechanism	# of Women using it (of 40 interviewed)
Restricted diet	32
Casual labor	25
Sold livestock	22
Gifts from family/friends/social networks	13
Horticulture	13
Relied on non-farm salaries	8
Sold chickens	6
'Economized'	5
Outside aid (govt. or private)	4
Irrigation	4
Have children work	4
Loans	3
Rely on savings	2
Sold milk	2
Sold local brew	2
Sold chapatti (a local bread) to casual laborers	2
Sold trees	2
Sold land	2
Sold eggs	1
Lease land	1
Sold matatu (a local term for taxi)	1
Bought food from far away, sold locally	1
Borrowed farming inputs	1
Fetch water at night	1
Sell household things	1
Dug a well	1
Reduced employees	1
Sold soap	1

drought caused crop failure, which exacerbated respondents' most fundamental vulnerability, lack of capital. Maize and beans are staple foods in Gituamba, and the inability of a household to grow them for consumption means they must be purchased. Alternatively, many households invested capital resources to re-plant crops after the first sowing failed, only to experience repeated crop failure and no economic return. Second, food insecurity caused respondents and their households to restrict their diets. Thirty-two of the respondents restricted their diets (Table 1), with most eating two small meals of ugali (boiled maize meal) and tea each day. This dietary restriction decreased the productivity of labor and human capital in affected households. Finally, the drought forced households to sell their livestock and, in a few cases, land (Table 1), or resulted in livestock death, all of which reduced households' productive assets. These three outcomes created potential poverty traps for over 50% of households.

Crop failure and food insecurity during the drought acted as exclusionary mechanisms that made it even more difficult for women to access badly needed capital. For respondents with alternative livelihood options, lack of access to capital was not enough, alone, to create a trap. However, depleted assets and lowered productivity of households often served as reinforcing dynamics, which minimized or eliminated secondary livelihood strategies that women usually employed. Figure 1 illustrates this dynamic.

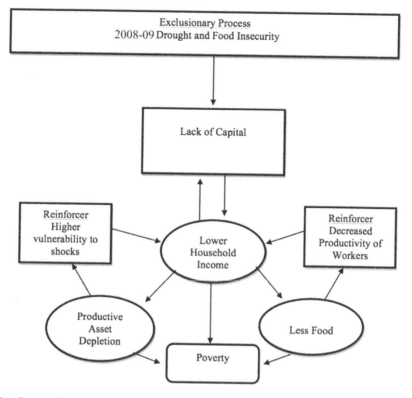

Figure 1. Poverty dynamics triggered by the 2008–2009 drought in Gituamba, Kenya.

During and immediately after the drought, many respondents lacked the capital needed for planting, so they planted less land than normal, planted late, or planted with even fewer inputs than normal. This resulted in lower than normal crop yields for the next harvest even though there was sufficient rainfall. Lower than normal yields may reinforce the poverty trap cycle for the coming year by making it difficult for a household to regain its pre-drought dietary and asset levels.

The creation of poverty traps was further reflected in a shift in many respondents' priorities during the 2008–2009 drought. When ranking drought-time priorities, 32 respondents said their first priority was acquiring food. Twelve respondents said that food was the only priority to which they could allocate resources during the drought. They spent so much of their time and resources securing food that they seldom fulfilled other obligations. The most common second priority was children: trying to keep them in school and maintaining their general well-being. Only five women listed a third priority; most women, in contrast, stated they had no resources or time leftover after addressing hunger and child welfare.

As the drought progressed, women experienced a fundamental shift away from non-drought priorities, such as improving livelihoods and increasing assets, towards survival activities like acquiring food and maintaining child welfare. This priority shift supports the idea that people who are focused on survival have very limited choices as to how they spend their time, energy and other resources (Nussbaum, 2000; Sen, 1999). This lack of meaningful choices is also an emergent characteristic of poverty traps.

Drought and women

Women of Gituamba faced other challenges during the 2008–2009 drought. As crop failure became imminent, over half of respondents sought more frequent casual labor opportunities or other off-farm work (Table 1), while still maintaining their normal household obligations. These same respondents reported they did not receive any additional help with household obligations because men were also out seeking work. Children were either in school or too weakened from an improper diet to contribute significantly to household tasks.

In addition to increased off-farm work, women had to walk further to find water, expend more effort to acquire food, and travel further to find casual employment as more and more farms became unproductive. The increased difficulty and number of obligations respondents took on, paired with an inadequate diet, resulted in a poverty of time and energy for many of them. Respondents who restricted their diet and relied heavily on casual labor as a coping mechanism struggled to manage tradeoffs between finding and purchasing food, working for low wages, and completing essential household tasks.

This poverty of time and energy also resulted in a change in community involvement. Nineteen of 32 women who described themselves as active in at least one or more community organizations said they were forced to drop community obligations during the drought. Preparing food, fetching water and caring for children are traditionally women's tasks in Gituamba. As their multiple, gender-specific, domestic obligations grew during the drought, so too did their poverty of time and energy for public involvement.

Other research has documented a similar poverty of time and energy among African women, particularly as they transition from the domestic sphere to the public sphere. May et al. (1998) found that, when South African women gained access to (or found it necessary to seek) outside work, they often received no help with existing household obligations. Jobs they were able to obtain outside the home were typically low paying, resulting in marginal economic gains. Women in Gituamba also fit this model. Those working outside the home usually earned very low wages as casual laborers on horticulture farms.

Coping mechanisms, resiliency and adaptation

Women employed many coping mechanisms to minimize and manage the effects of the 2008–2009 drought. In total, 28 unique strategies were used (Table 1). On average, each woman (or their household in general) employed four coping strategies. The three most commonly used mechanisms were dietary restriction, casual labor, and selling livestock, all of which were employed to manage food insecurity.

Although many respondents engaged in casual labor in normal years, they often did it sporadically to earn money for school fees or other expenses. In 2008–2009, it became a major source of income for 25 households, who used it to raise money for purchasing food, rather than non-essential expenses. Unfortunately, wages paid to casual workers during the drought decreased from about 200 shillings per day to 50 shillings per day. This was a result of the increased number of people looking for work.

In years of good rainfall, Gituamba is a very productive region. As a result, the community has become over-reliant on staple crop and livestock agriculture. When the rains fail, there are few options for maintaining food security and nutrition levels. Many households sold livestock during the drought as a means to purchase staple foods. Households sold 20% of their livestock holdings; another 28% of livestock died in 2008–2009. Food security is clearly Gituamba's greatest vulnerability. Addressing the community's over-reliance on staple crop and livestock agriculture will be paramount to successful adaptation.

Thirteen of the 40 households interviewed diversified their agricultural activities during the drought by participating in horticulture activities. Most of these households grew tomatoes, and had at least a bit of experience with horticulture from previous years. Four of the horticultural operations were watered using pipe irrigation systems. Nine respondents carried water to their horticulture fields. Those with pipe-irrigation harvested higher yields earlier in the season than those who hand-watered. Those with pipe-irrigated operations were therefore able to sell many tomatoes at a good price. Those with hand-watered operations harvested lower amounts, and did so later in the season. At their time of harvest, the market was flooded with tomatoes, so the price dropped significantly. This caused some of the respondents to sell their tomatoes at a loss, which was not an effective coping strategy. This demonstrates that introducing new, non-staple crop agricultural initiatives into Gituamba as a food security or income initiative may not be successful without improved supply management and price-risk management tools.

Seventeen women began selling various household and farm goods as a coping mechanism in 2008–2009. Products included homemade soap, chickens, milk, local brew, chapatti, tree seedlings and eggs. Income generated by the sale of these products was generally very low, but it was sufficient to justify continued production. The biggest challenge to this coping strategy was finding customers; many people could not afford even basic necessities. To get their products to non-drought areas, women needed capital for transportation, which they did not possess.

Another common coping mechanism included receiving help (money or food) from friends or family, primarily those living outside the drought-affected area. Adaptation policy should therefore consider how social networks and inter-community ties might be strengthened to reduce vulnerability. Even within Gituamba there is socio-economic stratification and intra-community social networks that could be drawn upon during crises. There is potential to strengthen such networks and use them for vulnerability reduction.

Little international aid was available to Gituamba in 2008–2009, in part because the region has a reputation for being fertile and productive. Only four respondents reported receiving food aid. When asked why they did not use food aid as a coping strategy, respondents said it was difficult to obtain due to a corrupt and highly-politicized distribution process. Even if they waited in line all day, they were not certain to receive aid. The high opportunity cost of their time outweighed the one or two kilograms of maize flour they might (or might not) receive.

Coping mechanisms not only help identify sources of vulnerability within a community, they also highlight areas of strength. Nine households displayed high resilience during the 2008–2009 drought. None of these nine reported any change in school attendance or health. Eight reported no change in diet, and one said they suffered only slight cutbacks in the amount of food eaten and experienced no change in the type of food they ate. All nine respondents reported no other areas of household well-being that suffered significantly during the drought.

These households were resilient due to a number of different factors. Four households were resilient due to their knowledge of irrigated agriculture and the ability to acquire necessary inputs. Formal employment was essential to the resiliency of three households, but was supplemented with productive assets and savings. One respondent was resilient due to successful business operations, market forces, and her access to credit. The final resilient household used social networks, combined with overall livelihood diversification, to increase her resiliency to drought. Future adaptation policies should consider and attempt to replicate these successful strategies: improved agricultural knowledge and infrastructure, formal employment, access to credit, and livelihood diversification.

Adaptation policy

The IPCC (2007) predicts a 20% reduction in growing season length in much of Kenya by 2050, which will put the food security of numerous communities at risk. Adaptation policy should aim to reduce vulnerability, yet also include a plan for relief operations in case they become necessary. As such, policy should combine relief, recovery, and long-term adaptation into a comprehensive plan.

Relief

Food insecurity was clearly the biggest challenge during the 2008–2009 drought. There was a severe lack of food aid in Gituamba, and a general perception that aid disbursement was riddled with corruption. Food aid may have controversial effects on local markets and farmers, but Gituamba faced a food deficit. Channeling food into the area through free aid, local markets, or food-for-work programs will be imperative during future droughts of similar severity. Reducing corruption in food aid distribution could also improve food security.

In addition to food insecurity, women faced a severe lack of time and energy during the 2008–2009 drought. Work programs that provide women a fair wage or food could help alleviate this by providing a reliable place for them to seek work. Rehabilitation of several large, communal boreholes that were originally dug in the colonial era could also alleviate women's poverty of time and energy. According to respondents, these deep boreholes would likely still contain water during a drought as severe as the one in 2008–2009. If successful, the rehabilitation of these boreholes would reduce the distance many residents have to travel to find a reliable water source, thereby making water collection a less time and energy intensive activity, particularly during drought. The Gituamba Umbrella Community Based Organization, discussed below, is working towards this goal.

One more issue to be considered when discussing relief is the depletion of livestock that most households experienced in 2008–2009. Livestock holdings in the area were reduced by 48%, either because of intentional liquidation (20%), or more troublesome, death before animals could be sold or slaughtered (28%). Death of livestock results in no gains, but rather overall net losses, for a household. A livestock-purchasing program during drought might therefore prove beneficial for many residents of Gituamba.

Recovery

The most essential piece of recovery policy is to ensure that women have enough access to capital and farming inputs to properly prepare for future growing seasons after a drought subsides. Microloans for certified seed and fertilizer could be an effective strategy for achieving this goal. These small loans must have low interest rates, however, and repayment plans that do not force households to sell their goods immediately after harvest, when prices are depressed. Ensuring that women have access to capital, farm inputs, and price-risk management options will help guarantee good yields in years following a drought. A good harvest will interrupt the creation of poverty traps, strengthen human capital by improving diets, jump start secondary income generation activities, and help reclaim productive assets lost during a drought. These loans could be made available in both normal and recovery years, as a longer-term strategy to address women's lack of capital. Research suggests that this type of microfinance is most successful when paired with training on proper use of farming inputs and improved agricultural practices (Ali-Olubandwa, Kathuri, Odero-Wanga, & Shivoga, 2011).

Long-term adaptation

To reduce vulnerability to future droughts, adaptation policy should find ways to improve households' access to capital. Access to business-related credit was instrumental in the success of at least one household that was resilient to drought in 2008–2009. Furthermore, many women expressed a desire to start small businesses during good years, but stated they did not have access to the necessary capital to cover start-up costs. Making microloans available to women interested in starting small businesses may be a successful approach (Matin & Hulme, 2003; Swain, Sanh, & Tuan, 2008).

Formation of cooperatives for selling milk, honey or horticulture crops could also prove beneficial. While improved transportation infrastructure would certainly help women sell such products, major improvements to roads and systems are unlikely in the near future. The creation of cooperatives, comprised of small-scale producers, may help overcome transportation and market barriers. By joining forces, producers would have a larger supply of the good, and thus gain bargaining power with buyers or processors. Large-scale buyers might be willing to travel to a central location in Gituamba to purchase and pick-up these goods, if available in sufficiently large quantities. This would eliminate individual households' logistical problem of transporting their goods to market. Alternatively, producers could spread the costs of transportation among many people, rather than each person being responsible for their own transportation costs.

Cooperatives could also facilitate livelihood diversification within the community, which would reduce overall vulnerability to drought. Irrigated horticulture and bee-keeping contributed to drought resiliency in 2008–2009. Development of these activities in non-drought years could empower women with built-in options for coping with drought, rather than scrambling to find casual labor and selling household assets.

In response to the 2008–2009 drought, residents created the Gituamba Umbrella Community Based Organization (GUCBO) to reduce drought vulnerability by forming a network between existing community self-help groups. A potential project for GUCBO is to use these self-help groups to establish a community grain bank. Community grain banks entail either a start-up donation of grain by an outside actor (government or Non-governmental Organization [NGO]), or they require members to make deposits of surplus grains during productive years to slowly build a store. The grain can then be consumed in times of drought or other hardship and later replenished by the consumers. An established community store could supplement members' diets during drought, which would improve not just food security, but also household productivity. Social networks were instrumental in creating resiliency in 2008–2009; a community grain bank developed and managed by GUCBO could potentially use these networks to further strengthen resiliency.

Finally, households that were resilient during the 2008–2009 drought demonstrated that irrigation has the potential to reduce vulnerability. Its effectiveness depends, however, on water availability. Improper irrigation by a few large horticulture operations could aggravate a water shortage during drought. Efficiently-designed systems for small-scale horticulture operations, in contrast, could increase households' livelihood options during times of staple-crop failure. Residents of Gituamba believe a successful irrigation system could be developed by rehabilitating a nearby groundwater source known as Kahiga Spring. A hydrological survey was beyond the scope of this study, but this option should be explored. Any irrigation initiatives should bear in mind that, under climate change, water may become even scarcer in Gituamba; in which case, irrigation could cease to be a viable adaptation strategy.

Adaptation ideas for further research

In addition to irrigated horticulture and forming cooperatives, there are several other agricultural strategies that might be successful at reducing vulnerability to drought in Gituamba. They include crop insurance, conservation agriculture, and diversifying agricultural activities to include cultivation of drought-resistant, rain-fed crops. Although these strategies were not in use in Gituamba at the time of this study, secondary research indicates they could hold potential. Each of these ideas would require further analysis and testing prior to implementation.

Conservation agriculture advocates reduced tillage, year-round soil cover, and rotational cropping (Hobbs, 2007). By eliminating the annual use of a plow, conservation agriculture reduces soil compaction and improves water infiltration. Research shows that this can mitigate water scarcity issues and reduce the likelihood of complete crop failure during drought (Hobbs, 2007; Rockstrom, Kaumbutho, Mwalley, & Temesgen, 2003). Eliminating plowing might also reduce capital requirements for farming; after all, most farmers in Gituamba pay someone to plow their land each year, or are left to do it by hand. Reducing soil disturbance and providing year-round soil cover would decrease erosion, thereby increasing organic matter and soil quality. Prior research has shown that increased infiltration, decreased erosion, and improved soil quality can result in higher crop yields, and in some cases allow for more crops to be planted throughout the year (Garcia-Torres, Benites, Martinez-Vilela, & Holgado-Cabrera, 2003; Hobbs, 2007). Yet, conservation agriculture also has barriers to entry, including increased labor for weeding or increased spending on herbicides to control weeds (Giller, Witter, Corbeels, & Tittonell, 2009). Such tradeoffs should be properly assessed before being promoted in Gituamba.

Crop and livestock insurance for small-scale farmers is just starting to become available in Kenya (Chantarat, Mude, Barrett, & Carter, 2013). For Gituamba, precipitation-based policies that pay out according to rainfall might be suitable. From a household's perspective, insurance policies would ideally compensate not only for the cost of farming inputs invested during an unsuccessful growing season, but also the estimated value of failed crops or livestock. It could be challenging to design a policy that fulfills this role, and is still affordable to farmers. It might also be challenging for households to save enough money to pay annual premiums during both non-drought and drought years (Chantarat, Mude, Barrett, & Carter, 2013; Cole, Bastian, Vyas, Wendel, & Stein, 2012). Index-based products are often used to overcome the problem of high premiums, but weather-based index products require a reasonable amount of investment in monitoring infrastructure (i.e., weather stations) from the insurer.

Rain-fed crop diversification should also be considered as an adaptation strategy. Diverse crops, even when planted at the same time, will be affected differently by precipitation patterns; this should help decrease the risk of complete crop failure (Adger, Huq, Brown, Conway, & Hulme, 2003). Incorporating drought tolerant crops, in particular, such as sorghum, millet, cowpeas and lentils, alongside traditional maize agriculture, could significantly reduce vulnerability to drought. Diversified agricultural systems provide additional benefits by lowering the risk of loss due to pest or disease outbreaks and increasing nutritional potential. However, there are barriers to diversification, including lack of knowledge about secondary crops, and lack of market incentives (Lin, 2011). As evidenced by the tomato surplus in Gituamba during drought, any crop diversification initiatives will need to be paired with strategies that increase market access and decrease price risk.

Conclusion

Adaptation to drought and climate change is likely to require a combination of household, community and, in some cases, third-party initiatives to be successful. Understanding a household's or community's sources of vulnerability, during both non-drought and drought years, is a useful means to inform adaptation policies. Vulnerability studies should aim to identify groups that are disproportionately affected by drought, recognize conditions that may cause them to enter a poverty trap, and understand options and barriers for coping effectively and escaping a drought-induced poverty trap. Comprehensive policies that involve relief, recovery, and long-term adaptation strategies are needed currently, and will become increasingly important in the face of climate change.

Acknowledgements

This author wishes to thank Dr Jean Garrison, the Kinyanjui family and the women of Gituamba for their help in developing this project from start to finish. This research was funded by several entities at the University of Wyoming including the Social Justice Research Center, the Arts and Sciences/Saunders-Walter Scholarship, the Dick and Lynne Cheney Scholarship and the Haub School for the Environment.

References

Adams, A. M., Cekan, J., & Sauerborn, R. (1998). Towards a conceptual framework of household coping: Reflections from rural West Africa. *Africa, 68*, 263–283.

Adger, W. N., Dessai, S., Goulden, M., Hulme, M., Lorenzoni, I., Nelson, D. R., . . . Wreford, A. (2007). Are there social limits to adaptation to climate change? *Climatic Change, 93*, 335–354.

Adger, W. N., Huq, S., Brown, K., Conway, & Hulme, M. (2003). Adaptation to climate change in the developing world. *Progress in Development Studies, 3*, 179–195.

African Partnership Forum. (2007). *Climate Change and Africa* (Working Paper). African Partnership Forum and the secretariat of the New Partnership for Africa's Development. Retrieved from http://www.africapartnershipforum.org/dataoecd/57/7/38897900.pdf

Ali-Olubandwa, A. M., Kathuri, N. J., Odero-Wanga, D., & Shivoga, W. A. (2011). Challenges facing small scale maize farmers in western province of Kenya in the agricultural reform era. *American Journal of Experimental Agriculture, 7*, 466–476.

Azaradis, C. (2006). Poverty trap theory: What have we learned. In S. Bowles, S. N. Durlauf, & K. Hoff, (Eds.), *Poverty Traps* (pp. 17–40). Princeton, NJ: Princeton University Press.

Barrett, C., Marenya, P., McPeak, J., Minten, B., Murithi, F., Oluoch-Kosura, W., . . . Wangila, J. (2006). Welfare dynamics in rural Kenya and Madagascar. *Journal of Development Studies, 42*, 248–277.

Bhattamishra, R. & Barrett, C. (2010). Community-based risk management arrangements: A review. *World Development, 38*, 923–932.

Brock, K. & McGee, R. (2002). *Knowing poverty: Critical reflections on participatory research and policy*. London: Earthscan.

Chambers, R. (1994). Participatory rural appraisal (PRA): Challenges, potentials and paradigm. *World Development, 22*, 1437–1454.

Chambers, R. (1995). Poverty and livelihoods: Whose reality counts. *Environment and Urbanization, 7*, 173–204.

Chambers, R. (1989). Vulnerability: How the poor cope. Sussex: University of Sussex.

Chantarat, S., Mude, A. G., Barrett, C. B., & Carter, M. R. (2013). Designing index-based livestock insurance for managing asset risk in northern Kenya. *The Journal of Risk and Insurance, 80*, 205–237.

Cole, S., Bastian, G., Vyas, S., Wendel, C., & Stein, D. (2012). *The effectiveness of index-based micro-insurance in helping smallholders manage weather related risks*. London: EPPI-Centre, Social Science Research Unit, Institute of Education, University of London.

Eriksen, S., Brown, K., & Mick, K. (2005). The dynamics of vulnerability: Locating coping strategies in Kenya and Tanzania. *Geographical Journal, 171*, 287–305.

Eriksen, S. & O'Brien, K. (2007). Vulnerability, poverty and the need for sustainable adaptation measures. *Climate Policy, 7*, 337–352.

Garcia-Torres, L., Benites, J., Martinez-Vilela, A., & Holgado-Cabrera, A. (2003). *Conservation agriculture*. Dordrecht, Netherlands: Kluwer Academic Publishers.

Giller, K. E., Witter, E., Corbeels, M., & Tittonell, P. (2009). Conservation agriculture and smallholder farming in Africa: The heretics' view. *Field Crops Research, 114*, 23–34.

Hobbs, P. R. (2007). Conservation agriculture: What is it and why is it important for future sustainable food production? *Journal of Agricultural Science, 145*, 127–137.

Intergovernmental Panel on Climate Change. (2007). *Climate Change 2007: Synthesis report*. World Meteorological Organization and United Nations Environmental Programme, Geneva. Retrieved from http://www.ipcc.ch/publications_and_data/publications_ipcc_fourth_assessment_report_synthesis_report.htm

Kaguara, A., Beethoven, M., Matere, J., & Koskei, T. (2009). *Laikipia district long rains food security assessment report*. Retrieved from http://www.kenyafoodsecurity.org/longrains09/district_reports/laikipia.pdf

Kamau, C., Anyango, G., Gitahi, M., Wainaina, M., & Downing, T. (1989). Case studies of drought impacts and responses in Central and Eastern Kenya. In T. Downing, K. Gitu, & C. Kamau (Eds.), *Coping with drought in Kenya* (pp. 211–230). Boulder, CO: Lynne Rienner.

Kenya Red Cross Society. (2009). *Drought appeal 2009: Alleviating human suffering*. Retrieved from http://www.reliefweb.int/rw/RWFiles2009.nsf/FilesByRWDocUnidFilename/MCOT-7NHF4S-full_report.pdf/File/full_report.pdf

Kumssa, A. & Jones, J. F. (2010). Climate change and human security in Africa. *International Journal of Sustainable Development and World Ecology, 17*, 453–461.

Lin, B. B. (2011). Resilience in agriculture through crop diversification: Adaptive management for environmental change. *BioScience, 61*, 183–193.

Little, P., Stone, M., Mogues, T., Castro, A., & Negatu, W. (2006). Moving in place: Drought and poverty dynamics in South Wollo, Ethiopia. *Journal of Development Studies, 42*, 200–225.

Liverman, D. (2011). *Informing adapation* (World Resources Report). Retrieved from http://www.worldresourcesreport.org/responses/informing-adaptation

Matin, I. & Hulme, D. (2003). Programs for the poorest: Learning from the IGVGD program in Bangladesh. *World Development, 31*, 647–665.

May, J., Govender, J., Budlender, D., Renosi, M., Rogerson, C., & Stavrou, A. (1998). Poverty and Inequality in South Africa. Report prepared for the Office of the Executive Deputy President and the Inter-Ministerial Committee for Poverty and Inequality. Retrieved from http://www.polity.org.za/polity/govdocs/reports/poverty.html.

Narayan, D., Chambers, R., Shah, M., & Petesch, P. (1999). *Global synthesis: Consultations with the poor*. Retrieved from http://siteresources.worldbank.org/INTPOVERTY/Resources/335642-1124115102975/1555199-1124138742310/synthes.pdf

Narayan, D., Patel, R., Schafft, K., Rademacher, A., & Koch-Schulte, S. (1999). Can anyone hear us? In *Voices of the Poor*. New York, NY: Published for the World Bank, Oxford University Press.

Nussbaum, M. (2000). *Women and human development*. Cambridge: Cambridge University Press.

Rockstrom, J., Kaumbutho, P., Mwalley, P., & Temesgen, M. (2003). Conservation farming among small-holder farmers in E. Africa: Adapting and adopting innovative land management options. In L. Garcia-Torres, J. Benites, A. Martínez-Vilela, & A. Holgado-Cabrera (Eds.), *Conservation agriculture* (pp. 459–469). Dordrecht, Netherlands: Kluwer Academic Publishers.

Sachs, J. (2005). *The end of poverty: Economic possibilities for our time*. New York, NY: Penguin Books.

Sen, A. (1999). *Development as freedom*. New York, NY: Anchor Books.

Smucker, T. & Wisner, B. (2008). *Changing household responses to drought in Tharaka, Kenya: Vulnerability, persistence and challenge*. Oxford, UK: Blackwell.

Solomon, S., Qin, D., Manning, M., Marquis, M., Averyt, K., Tignor, M., Miller, H.L., & Chen, Z. (2007). Climate Change 2007: Working Group I: The Physical Science Basis. Cambridge: Cambridge University Press.

Speranza, C., Kiteme, B., Ambenje, P., Wiesmann, U., & Makali, S. (2010). Indigenous knowledge related to climate variability and change: Insights from droughts in semi-arid areas of former Makueni District, Kenya. *Climatic Change, 100*, 295–315.

Swain, R. B., Sanh, N. V., & Tuan, V. V. (2008). Microfinance and poverty reduction in the Mekong Delta in Vietnam. *African and Asian Studies, 7*, 191–215.

World Food Programme. (2009). *WFP seeks urgent assistance as Kenya sinks deeper into crisis.* Retrieved from http://www.wfp.org/news/news-release/wfp-seeks-urgent-assistance-kenya-sinks-deeper-crisis

Yohe, G. & Tol, R. (2002). Indicators for social and economic coping capacity – moving toward a working definition of adaptive capacity. *Global Environmental Change – Human and Policy Dimensions, 12,* 25–40.

Medium-term electricity load forecasting and climate change in arid cities

Bhagyam Chandrasekharan[a] and Bonnie Colby[b]

[a]Department of Agricultural Economics, Washington State University; [b]Department of Agricultural Economics, University of Arizona

Electric utilities need to consider how potential changes in climate patterns will affect their peak loads. This study incorporates weather and socio-economic variables into a medium-term load forecasting model to consider potential climate change effects on the challenging summer peak season for utilities in the arid southwestern US. Our 'average hourly load by month' model shows marked improvement over a purely autoregressive approach to load forecasting used by some electric utilities. In light of climate change, electric utilities and society can benefit from minimizing inaccuracies in load predictions. Decision-making based on more climate-sensitive forecasts will reduce the water and carbon footprint of electric utilities and improve their investment strategies for renewable energy technologies.

Introduction

Climate and weather are major causes of variation in demand for electricity. The sensitivity of energy demand to weather stems from the fact that electricity generated must be instantly consumed (Psiloglou et al., 2009). Climate change affects electricity consumption (load), which in turn influences demand for water and changes in air quality associated with electricity generation. The increased temperatures and variability in precipitation associated with climate change are a growing challenge for electric utilities. Growing scarcity of water supplies affect both electricity demand and generation in the arid southwestern US (Electric Power Research Institute, 2006). In the desert regions of Nevada, Arizona and southern California, which are characterized by low rainfall and high summer heat, peak electricity demand and peak water demand both occur in the summer months.

Electricity load forecasting is an integral part of electric utilities' planning. Accurate forecasts mitigate costs associated with load switching, overloading, blackouts and equipment failures. Electric load varies with time of day, day of week, season, climatic conditions, and past usage patterns (Alfares & Nazeeruddin, 2002). With climate change, weather variations are expected to exhibit a pattern distinct from the past. Efficient production by electric utilities, based on improved load forecasts, will reduce their carbon footprint, a key concern given only 11% of US electricity is generated from renewable sources (US Energy Information Administration [USEIA], 2010).

This study investigates the potential effects of climate change on electricity demand using medium-term load forecasting models. We develop a forecasting model for average hourly load by month, one month in advance. This time frame is critical for operational planning, for ensuring supply reliability, for making capital investment decisions, and for evaluating energy price contracts.

Our monthly load model for the Tucson Metropolitan Statistical Area (MSA, see Abbreviations list at end of article) is based on load data from the largest electric utility in the area. We evaluate our forecast using generally accepted measures of forecasting error; we also estimate utility cost savings due to improved forecasts.

Electricity load and load-forecasting

A shortage in electricity supply can lead to brownouts, blackouts and financial losses incurred through purchases in electricity spot markets. Failure to provide adequate power for space cooling during summer heat can also contribute to mortality in segments of the population most vulnerable to heat stress. Overproduction by an electric utility, in contrast, implies wastage of scarce resources and higher production costs. Because commercially viable storage technology for electricity is lacking, load forecasting is crucial to assure an uninterrupted, secure and affordable supply of electricity. Electric load forecasting involves predicting magnitudes of electric load (i.e., demand) over different time periods (hours, months) for specific locations and spatial scales (Alfares & Nazeeruddin, 2002).

Electric utilities are most concerned with peak load forecasts, that is, the maximum instantaneous load or the maximum load over a designated interval of time. Peak load forecasts can be made at the daily, monthly, seasonal or annual level. Monthly and annual peak load forecasts are important for securing adequate generation, transmission and distribution capacities. These medium-term peak load forecasts improve capital expenditure decisions and electric system reliability (Feinberg, 2009).

The electricity industry was characterized, in the past, by a highly vertically-integrated market structure with little competition. This structure has been replaced by competitive markets, which have increased the costs of over or under-production, as well as the selling or buying of electricity on the spot market. Electricity for an urban area is typically produced by several generating units with different lead times. Load forecasts therefore allow each unit to more accurately anticipate its usage, and avoid costs incurred from idle operation. Load forecasting is an integral process for electric utilities, energy suppliers, system operators and other market participants. Because the financial penalties for forecast errors are so high, reducing error by a fraction of 1% is a meaningful improvement (Weron, 2006).

Impact of climate change

Climate assessments predict that the US southwestern desert region will experience continued temperature increases, summer heat waves that are longer and hotter than experienced in the past, and reduced stream flows, among other climate and ecosystem changes (Garfin, 2012). By the end of the century, average annual temperature across the southwest US is projected to rise between 4 °F and 10 °F (roughly 2.2 °C and 5.6 °C) above the historical baseline (US Global Change Research Program [USGCRP], 2009). Compared to coarser-resolution Intergovernmental Panel on Climate Change (IPCC) climate models, downscaled climate models for the region project an additional 0.9 °F (0.5 °C) of warming and an additional 3 mm/month decline in precipitation (Dominguez, Canon, & Valdes,

2010). These downscaled projections imply the region will be 3.52 °C (or 6.33 °F) warmer and 16.25 mm drier by the year 2050 (Dominguez et al., 2010). Warmer temperatures and reduced water supplies will challenge energy supply reliability by increasing summer cooling demand and decreasing power generation efficiency (Garfin, 2012; USGCRP, 2009).

Continuing population growth in the Southwest will also intensify energy management challenges by increasing total energy demand and increasing competition for scarce water supplies. Water is an integral element of electricity generation; it is used directly in hydroelectric generation and indirectly for cooling and emissions scrubbing. Thermoelectric power generation accounts for nearly 40% of all freshwater withdrawals in the US (Hutson et al., 2004). As climate change and population growth increase competition for water, the energy sector will be strongly affected due to its large water requirements (USGCRP, 2009).

To complicate matters at the water-energy nexus, electricity is a primary input for treating and delivering water for drinking and irrigation. In the arid Southwest, peak demand for agricultural and municipal water use, with its accompanying energy requirements, occurs in the same hot months as peak energy demand for cooling (Garfin, 2012). By increasing energy demand for cooling in the summer, when energy demand is also highest for agricultural and municipal water users, climate change will significantly increase summer-season peak loads (USGCRP, 2009). During the winter, climate change induced warming might actually decrease demand for heating energy. However, heating during the winter is less important in the Southwest than space-cooling during the summer.

Electric utilities are very concerned with the effects of climate change because trends in electricity consumption show a direct relationship to temperature and precipitation. In 2005, electric utility company specialists and stakeholders who attended a workshop titled 'Identifying Research to Help Electric Companies Adapt to Climate Change', unanimously agreed that more applied research on climate change impacts was necessary. Load forecasting was one of the key areas considered for application.

Despite their enthusiasm for load-forecasting research, most industry-specific climate change reports have instead emphasized mitigation of greenhouse gas (GHG) emissions. Although mitigation is integral to the electric industry's response to climate change, so is adaptation. To support the industry's climate change adaptation efforts, it is necessary to understand and accurately predict electricity demand in light of climate change.

Past studies

Relationships between weather variables and electricity load have been widely investigated for load forecasting. Lam (1998) investigated the relationships between residential electricity consumption, economic variables and temperature for Hong Kong. They found that including temperature, measured as cooling degree days, facilitated estimation of residential electricity use. Ranjan and Jain (1999) modeled electricity consumption as a function of population and weather-sensitive parameters, specifically sunshine hours, temperature, rainfall and relative humidity. They created seasonal models and found that, in each of the seasons, different weather variables were relevant for explaining electricity consumption. Sailor and Muñoz (1997) developed models for eight US states, and found electricity consumption was related to climate parameters.

Temperature plays the most important role in electricity load (Psiloglou et al., 2009), due to the use of electricity for heating and cooling. Temperature exhibits a non-linear pattern in its relationship with load, often referred to as a 'U-shaped' or 'hockey-stick' curve. Figure 1 illustrates this concept for a southern Arizona electric utility, TRICO.

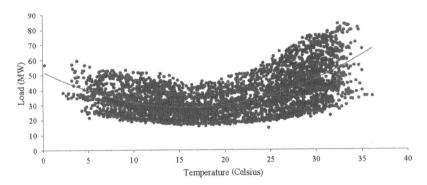

Figure 1. (Color online) Relationship between temperature and electricity load for the southern Arizona electric utility, TRICO.

In winter, a negative relationship exists between load and temperature because an increase in temperature diminishes the need for energy resources used for heating (Bessec & Fouquau, 2008). In summer, an increase in temperature raises the demand for electricity for cooling. At an intermediate or 'neutral' temperature range, load demand is not sensitive to temperature changes. The neutral temperature for Tucson MSA is around 59 °F (15 °C) (Tanimoto, 2008).

To model this non-linear relationship, prior studies have used degree-day variables (Al-Zayer & Al-Ibrahim, 1996; Amato, Ruth, Kirshen & Horwitz, 1996; Pardo, Meneu, & Valor, 2002; Sailor & Muñoz, 1997), logistic smooth threshold regression models (Bessec & Fouquau, 2008; Moral-Carcedo & Vicéns-Otero, 2005), and semi or non-parametric models (Engle, Granger, Rice, & Weiss, 1986; Henley & Peirson, 1997). Bessec and Fouquau (2008) use logistic smooth threshold regression (LSTR) model to capture the non-linearity of this relationship for 15 countries in the EU, using data for a period of 20 years.

Suganthi and Samuel (2012) provide a recent review of modeling approaches used in energy demand forecasting, including electricity load forecasts. They note increased emphasis on inclusion of climate and economic variables, such as cooling degree days, income, population, and household size. Hyndman and Fan (2010) use density forecasting to estimate probability distributions for future long-term electricity loads. They incorporate temperature, along with calendar effects, price changes and economic growth variables. Mirasgedis et al. (2006) develop models for medium-term electricity load forecasting that incorporate relative humidity, and heating and cooling degree-days. They note that electricity infrastructure planning can be improved by investigating cooler or warmer conditions expected for the future.

Precipitation also influences electricity demand. Specifically, an increase in precipitation decreases temperature, which decreases load demand for air-conditioning (Willis, 2002). While this suggests that load and precipitation would be inversely related, rainfall pushes people indoors, increasing their electricity consumption (Willis, 2002). The relationship between precipitation and load appears to be location-specific, with past studies finding both positive and negative effects of precipitation on load.

Relatively few studies examine the link between humidity and electricity consumption, but humidity increases the heat-retention capacity of air (Willis, 2002). This tends to increase the need for cooling during summer months, and thereby increase electricity load (Contaxi, Delkis, Kavatza, & Vournas, 2006). Rapid winds are known to generate a chilling sensation, which decreases the need for summer electric cooling but augments the need for heating during the winter. Finally, solar radiation has an ambiguous effect on load. Higher illumination of surroundings increases ambient temperature, thereby increasing the demand for cooling (Willis, 2002). However, an increase in solar illumination also decreases the use of electric lighting devices.

Despite a large body of literature on relationships between climate elements and electricity load, our study is one of the few to focus on the distinctive seasonal electricity-load patterns of the US arid southwest, where peak water demand and peak space cooling demand occur simultaneously over the summer months.

Existing load forecasting models

Several methods have been used to model and forecast the relationship between electricity load and various factors, such as the time of day, day of week, season, climatic conditions, and past usage patterns. Load forecasting models fall into the following categories:

(1) *Statistical models:* This category includes multiple regression models, exponential smoothing, adaptive load forecasting, and time series analysis. These models represent load as a function of factors such as time, weather, customer class, and past load values.
(2) *Computational intelligence techniques:* These highly-sophisticated techniques include neural networks, fuzzy logic and genetic algorithms.
(3) *Knowledge-based expert systems:* These systems employ the knowledge and analogical reasoning of experienced human operators. They use artificial intelligence concepts to emulate human performance.

Temporal aspects of load forecasting can be classified in terms of the planning horizon: up to one day for short-term load forecasting (STLF), one day to one year for medium-term load forecasting (MTLF), and one to 10 years for long-term load forecasting (LTLF) (Srinivasan & Lee, 1995). This paper seeks to incorporate climate information into the forecasting of average hourly load by month using statistical models. Our experience with local electric utilities in the arid Southwest indicates they customarily rely upon purely autoregressive (AR), autoregressive moving average (ARMA), or autoregressive integrated moving average (ARIMA) models for load forecasting. These predict future load based solely on past load data. ARIMA is an extension of ARMA to accommodate non-stationary processes (Weron, 2006). ARMA and ARIMA model forecasts are essentially extrapolations of a previous load history. The purpose our study is to explore how climate factors affect peak load for an arid city, so we develop a model that can incorporate climate information and has autoregressive errors.

Climate-adapted monthly load forecasting model
Data description

We create a medium-term model for predicting average hourly load by month for the Tucson MSA (Chandrasekharan, 2011). The load data consist of hourly measurements of total electricity for January 2000 through December 2009 for a large service area covering most of the Tucson MSA, including some copper mines. Tucson MSA data and Pima County data are used interchangeably because of their strong geographical overlap.

Table 1.   Summary statistics for the forecasting model of average hourly load by month.

Variable	Units	Mean[a]	Std Dev	Minimum	Maximum
Load	MW	1099.68	213.66	818.47	1593.08
Temp	°F	68.23	13.68	46.00	89.00
Precip	mm	0.03	0.04	0.00	0.18
RH	%	40.61	13.12	16.00	69.00
Windspd	m/s	4.11	0.65	2.50	5.50
SolRad	MJ/m²	482.39	143.99	247.00	754.00
Cuprod	Metric ton	12,005.58	2880.45	7640.32	17,913.83
PCY	$/person	29,708.63	3494.60	24,885.50	34,058.00
Popnch	Change in number of persons	1095.28	1538.53	−3570.76	2297.63

Note: [a] T = 120.

To control for seasonality, we include both weather and socio-economic explanatory variables in our regression model. Monthly average measurements for temperature, precipitation, relative humidity, solar radiation and wind speed data were gathered from the Arizona Meteorological Network (AZMET). Socio-economic data was acquired from a variety of sources. US Census annual population estimates were supplemented with estimates from the Arizona Department of Economic Security. Monthly population estimates were constructed using traffic count data from the Pima County Department of Transportation. Copper mine production data were obtained from the Arizona Department of Mines and Mineral Resources (AZDMMR) (Niemuth, 2004, 2005, 2006, 2007, 2008) and from the US Geological Survey (USGS). Annual per capita income data were sourced from the US Department of Commerce and the University of Arizona Economic and Business Research Center. Table 1 provides summary statistics for key variables in our model.

Dependent variable

The dependent variable for our model is average hourly electricity load by month for the Tucson MSA region, obtained by averaging load values measured at the hourly level for each month. In effect, the dependent variable is the average hourly load during a particular month of a specific year. Figure 2 shows that average hourly load by month exhibits considerable seasonality and has a gradually increasing trend. Seasonality at the monthly level is explained by the close relationship between electricity load and seasonal weather conditions. The load data show peaks during the monsoon months of July and August when cooling need is greatest, and relative troughs during the rest of the year.

Explanatory variables

To adapt medium-term load forecasting models to climate change, our model includes the following weather variables (whose coefficients' expected signs are reported in Table 2).

- *Temperature (Temp):* To model non-linearity of the relationship between load and temperature, we included a linear term (Temp), a squared term (Sqtemp), and an interaction variable with wind speed (Int2).

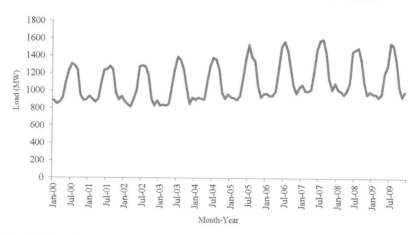

Figure 2. (Color online) Average hourly load by month for the Tucson Metropolitan Service Area.

Table 2. Expected signs of coefficients for the monthly model.

Variable	Description	Expected Sign
Temp	Temperature (°F)	Negative
Temp^2	Temperature squared	Positive
RH	Relative humidity	Negative
Solrad	Solar radiation	Positive
Precip	Precipitation	Negative
Windspd	Wind speed	Positive
PCY	Per capita income	Positive
Popnch	Population change	Positive
Cuprod	Copper mine production	Positive
Int2	Temp*Windspd	Negative

- *Precipitation (Precip):* Unlike hourly measurements of precipitation, which suffered from very small magnitudes, monthly average precipitation measurements were more substantial.
- *Relative humidity (RH)*
- *Solar radiation (Solrad)*
- *Wind speed (Windspd)*

To account for the seasonality of average hourly load by month, we include the following socio-economic variables in our analysis:

- *Change in monthly population (Popnch):* Population increases are generally associated with increases in electricity consumption by households, associated businesses and infrastructure, so we expect the monthly population variable to have a positive coefficient. In Tucson, monthly population shows significant variation over the calendar year due to the departure of college students in the summer and the arrival of vacationers during the mild winter. Annual US Census estimates are unable to capture this phenomenon adequately, so a monthly population estimate was created using traffic count data for Pima County. This assumes variations in traffic count are directly and positively related to changes in population.

- *Per capita income (PCY):* We expect that an increase in per capita income leads to increased purchase and use of power-consuming appliances. Therefore, we expect the coefficient associated with this term to be positive. Annual per capita income data were obtained from the US Census. This dataset was scaled down to the monthly level, assuming equal incremental increases in per capita income from month to month.
- *Copper mine production (Cuprod):* Due to confidentiality issues, the actual composition of the Tucson Electric Power (TEP) load data was not disclosed. However based on general information regarding TEP's clientele, it was evident that TEP provides electricity to three major copper mines in the region, namely, Mission Complex, Sierrita, and Silver Bell. Therefore, in lieu of specific data on electricity used by each mine, this study included total copper mine production data from the AZDMMR for these three mines. This dataset was annual, so monthly data were obtained through extrapolation, based on newspaper reports related to strikes and shutdowns at each of the mines. This allowed for sufficient month to month variation in this variable. We expect its coefficient to be positive.

Model estimation

To estimate the 'average hourly load by month' forecasting model using Ordinary Least Squares (OLS), it was necessary to test for heteroskedasticity using White's test (Table 3). No heteroskedasticity was found, a conclusion further supported by the Breusch-Pagan test (Table 3).

Based on a highly significant Durbin-Watson test statistic of 0.8947 (Table 4), it was evident that the errors exhibited autocorrelation. These findings were further supported by the Godfrey-Lagrange multiplier test (Table 5). Autocorrelation was anticipated not only because the model is based on a time series, but also because weather variables are correlated with each other.

Table 3. Tests of heteroskedasticity.

Test	Test Statistic	DF	Pr > ChiSq
White's Test	74.53	62	0.1322
Breusch-Pagan	13.91	11	0.1769

Table 4. Durbin-Watson test for autocorrelation.

Order	Durbin-Watson	PR < DW	Pr > DW
1	0.8947	<0.0001	1

Table 5. Godfrey-Lagrange Multiplier Test for autocorrelation.

Equation	Alternative	LM	Pr > LM
Load	1	35.50	<0.0001
	2	35.50	<0.0001
	3	38.24	<0.0001
	4	38.24	<0.0001

Due to the presence of autocorrelation, we used Feasible Generalized Least Squares (FGLS) for estimating a load forecasting model with autoregressive terms.

Analysis of results

The purpose of our study is to construct a forecasting model for average hourly load by month that allows us to examine potential effects of regional climate change on electricity loads. In the process of constructing this model, we also examine the econometric implications of incorporating weather variables and seasonality indicators into the purely autoregressive and ARIMA models that some local utilities use.

We first describe the criteria used to evaluate the performance of the new load forecasting models. We then present results for the models and briefly discuss the significance of coefficient estimates in each of the models. Finally, we generate forecasts for the next model year based on two hypothetical climate change scenarios.

Criteria for evaluating load forecasting models

Murphy (1993) recognized three types of 'goodness' of forecasts: consistency, quality, and economic value.

- *Consistency:* A forecast is said to be consistent if it corresponds to the forecaster's best judgement derived from their knowledge base (Murphy, 1993). We check the consistency of our load forecasts by determining if the signs of estimated coefficients in our load forecasting models are consistent with our expectations, which are based on an in-depth literature review.
- *Quality:* If a forecast corresponds closely to the observed values at (or during) the valid time of the forecast, it is generally considered to be of superior quality. This has been the primary focus of most forecasting studies, and is often calculated using measures of mean absolute error, mean-square error, and various skill scores. We use two familiar measures of forecast quality: Mean Absolute Percentage Error (MAPE), which is familiar to the electricity industry, and Root Mean Squared Error (RMSE), which is familiar to forecasters. MAPE is given by the following formula:

$$MAPE = \frac{1}{T} \sum_{t=1}^{T} \left| \frac{A_t - F_t}{A_t} \right| \tag{1}$$

where A_t is the actual value, F_t is the forecast value, and T is the size of the forecasting sample. This measure of error is unitless, and thus comparable across different models. It also assigns equal weights, in absolute terms, to both positive and negative errors. Note that MAPE will be undefined when the actual value is zero; however, electricity load data in our study do not include zero values.

Another relevant measure of error is the RMSE, calculated by the following formula:

$$RMSE = \sqrt{\frac{1}{T} \sum_{t=1}^{T} (A_t - F_t)^2} \tag{2}$$

Because errors are squared, RMSE gives greater weight to large errors, and is thus very sensitive to the presence of large errors. RMSE can also only be compared across models that are measured in the same units. MAPE and RMSE tend to move in tandem, that is, a 'good' (or 'bad') model will perform well (or 'poorly') in both MAPE and RMSE.

In addition to MAPE and RMSE, we also report Peak Load Error as a percentage, which measures our forecast's accuracy in predicting peak load during the given forecasting time period. It is calculated by the following formula:

$$Peak\ Load\ Error\ \% = \left(\frac{Actual\ Peak\ Load - Forecasted\ Peak\ Load}{Actual\ Peak\ Load} \right) \times 100 \quad (3)$$

- *Economic value:* Forecasts are said to have economic value if decision-makers real-ize incremental economic benefits through use of the forecasts. Electric utilities experience cost-savings through the use of more accurate forecasts by avoiding over and underproduction. Using International Energy Agency (IEA) estimates of costs of production under different generation technologies, we provide rough estimates of these cost-savings (see Appendix 1). Other benefits from using better forecasts include improved capacity planning and operational efficiency, which we do not attempt to quantify.

Additional load data for evaluating forecasts were not available, so we tested the accu-racy of our model forecasts through the use of 'hold-out' samples. With this technique, the time period of the load data used to fit the models ends before the end of the full data series. The remainder of the load data is retained for use in evaluation. With respect to the 'fit' period, the hold-out sample is a period in the future, used to measure the forecasting accu-racy of the models. We use the hold-out sample approach by keeping the year 2009 as our hold-out sample, and estimating our forecasting model using data up to December 2008.

It is important to acknowledge that we use actual weather data as explanatory vari-ables in our models, as opposed to daily or seasonal weather forecasts. The hold-out sample approach described above therefore assumes perfect knowledge of weather events. In reality, electric utilities use next-day or next-month weather forecasts to predict load. Therefore, the quality of weather forecasts used in forecasting is an issue to be taken into consideration. For the purposes of understanding how climate variables affect load in this region, we use actual weather data. We cannot assess the performance of our model if weather forecasts are used instead of observed weather data; however, our load forecast would presumably only be as good as the weather forecast used to generate it. Future work should investigate the use of regional weather forecasts to create an operational forecasting tool.

To explore the implications of regional climate change, we generate average hourly load forecasts by month for one year, based on two hypothetical climate scenarios:

(1) *Simple Change scenario:* In this scenario, following predicted impacts of climate change in the Southwest (Dominguez et al., 2010), temperature is increased at a steady rate equivalent to a 3.52 °C (or 6.33 °F) increase over the period 2010 to 2050. More specifically, we gradually increase temperature for each month in the calendar year as compared to the identical month in the previous year. Regional climate models' predictions for future precipitation are highly variable. Here, for illustrative purposes, we simply diminish each month's precipitation by 10% as compared to the identical month in the previous year.

(2) *More Intense Summer Shift (MISS) scenario:* This scenario assumes a 50% decline in monsoon precipitation, a temperature increase of 7.2 °F (4 °C) during the summer months from May to August, and an increase of 4.5 °F (2.5 °C) during the other months of the year. All other factors are held constant between the MISS and Simple Change scenarios.

Load forecasting model results

Table 6 shows the monthly model coefficient estimates with their associated standard deviations. The coefficients of all variables, except solar radiation and relative humidity, are significant ($p < 0.05$). With the exception of relative humidity and precipitation, all other variables show the expected signs. The positive coefficient on precipitation lends credence to the theory that rainfall in Tucson pushes people indoors, increasing their consumption

Table 6.   Monthly model Yule-Walker estimates.

Variable	Full Model	Basic Model
Intercept	1797***	1099***
	(121.323)	(23.0153)
Temp	−62.907***	
	(2.864)	
Sqtemp	0.601***	
	(0.024)	
RH	0.729	
	(0.374)	
SolRad	0.002	
	(0.053)	
Precip	184.390*	
	(75.140)	
Windspd	75.384**	
	(24.488)	
Int2	−1.294**	
	(0.355)	
PCY	0.022***	
	(0.001)	
Popnch	0.007**	
	(0.002)	
Cuprod	0.004**	
	(0.001)	
Lag1	−0.456**	−0.585**
	(0.085)	(0.06)
Lag3	0.188**	0.262**
	(0.083)	(0.042)
Lag12	−0.236**	−0.446**
	(0.083)	(0.055)
AIC	1,077.22	1,364.44
SBC	1,116.13	1,375.59
Forecasts for 2009		
RMSE	34.80	94.24
MAPE	2.80	7.33

Note: Significance levels: * $= 0.05$, ** $= 0.01$, *** $= 0.001$

of electricity. The generally-accepted cooling effect of precipitation does not seem to be valid for the Tucson MSA.

Per capita income and change in population coefficients are positive and significant, implying that load demand increases with population and per capita income. The positive and highly significant coefficient associated with copper mine production is consistent with the fact that Sierrita, Mission Complex and Silver Bell mines are significant contributors to load demand.

The highly significant autoregressive lags of order 1, 3 and 12 were included in the monthly model to account for autocorrelation. The coefficients for these terms indicate that load in the current month is positively related to load in the previous season (lag3), but electric load in the previous month and the previous year have a negative impact on current month's load. This is an unexpected result which merits deeper investigation in future studies.

From Table 6 we can clearly see that the Full model, which uses weather and socio-economic variables, performs better in terms of RMSE and MAPE of 2009 load forecasts than the purely autoregressive model (referred to as the Basic model) that some electric utilities use. On average, Full model forecasts are 4.73% more accurate than Basic model forecasts.

Based on per-unit generation cost estimates provided in the study, 'Powering Arizona' (Considine & McLaren, 2008), cost-savings from using the Full model as opposed to the Basic model are substantial (Table 7). This shows the economic value of improved load forecasts by incorporating weather information (albeit perfect weather information, which admittedly is unrealistic). To calculate economic value, we consider the following two situations in the electricity spot market:

- *Case A:* Penalty for purchasing electricity in the spot market is equal to the cost of generating a unit of electricity.
- *Case B:* Penalty for purchasing electricity in the spot market is equal to twice the cost of generating units of electricity, that is, losses are asymmetrical.

Here, the 'penalty' in the spot market refers to the spot market price for electricity. Without reliable data for our region about the cost of over or under-production, we assume the above two cases. These cases provide a simple explanation of the monetary value of improved forecasts. Future work should study the impact of asymmetric losses in the electricity spot market on the economic value of forecasts.

Figure 3 illustrates load demand response under the Simple Change scenario. The forecasted peak load, under simple climate change, is approximately 300 MW higher than for 2009. Other than a clear spike in forecasted load in the month of January, forecasted load

Table 7. Annual cost savings from improved monthly forecasts for 2009.

Technology	Average Cost of Generation[a/] ($/MWh)	Spot Market Penalty[b/] ($/MWh)	Case A: Annual Cost Savings ($)	Case B: Annual Cost Savings ($)
Scrubbed Coal	51.5	103	23, 802, 647.86	42, 249, 940.55
Conventional Gas	106	212	48, 991, 857.74	86, 961, 042.68

Note: [a/] 2008: Powering Arizona by Timothy Considine.
[b/] Spot Market Penalty = 2 × Average Cost of Generation.

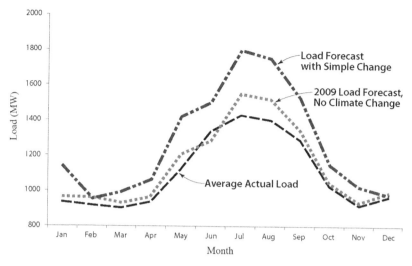

Figure 3.   Simple Change scenario for the monthly model.

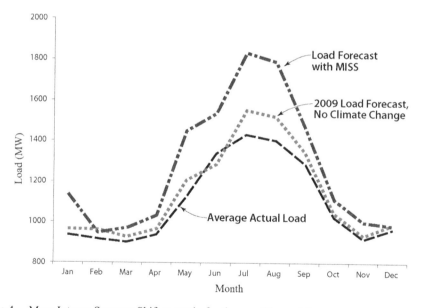

Figure 4.   More Intense Summer Shift scenario for the monthly model.

appears to follow a similar seasonal pattern as load in 2009 and average load across years 2000 to 2008.

In Figure 4, load forecasts for the MISS scenario show the expected load response when temperatures and precipitation change dramatically. Forecasted annual peak load under the MISS scenario, while still occurring in July, is more pronounced and is greater than load under the Simple Change scenario in the same month in 2009. It must be noted that the seasonal patterns remain intact in either of the climate change scenarios. Moreover, in both scenarios, load demand at the beginning of the year shows an unexpected spike. However, in MISS, the load curve appears to be more highly peaked in July, as compared to the Simple Change scenario.

We also calculate marginal effects on loads, using the mean values for significant explanatory variables. Because the coefficients for relative humidity and solar radiation were not significant, their marginal effects were not calculated. Due to the presence of autoregressive terms at lags 1, 3 and 12, marginal effects of significant explanatory variables in the monthly model were estimated using the following steps:

(1) Estimate the long-term value of the dependent variable, average hourly load in a specific month (Z_t), using the following formula:

$$Long\ Term\ Value\ of\ Z_t = \left(\frac{\hat{\mu} + \sum_{i=1}^{10} \hat{\beta}_i \hat{X}_i}{1 - \widehat{\emptyset}_1 - \widehat{\emptyset}_3 - \widehat{\emptyset}_{12}} \right) \tag{4}$$

where i = 1, 2 . . . .10 is the number of explanatory variables and t = −12, −11,.., 0,..,1, 2, 107 refers to a specific month in the time period of 120 months. $\hat{\mu}$ is the intercept term estimate from the model results and $\widehat{\emptyset}_1$, $\widehat{\emptyset}_3$, and $\widehat{\emptyset}_{12}$ are estimated model coefficients for the AR(1), AR(3) and AR(12) terms. $\hat{\beta}_i$ refers to the model coefficient for explanatory variable X_i

(2) Assume that the estimated value of the dependent variable $\left(\hat{Z}_t\right)$ for the first 12 time periods (i.e., first 12 months) is equal to the long-term value calculated in the previous step. The estimated value of the dependent variable for the remaining time periods (i.e., from t = 0 to t = 107) is then calculated using the following formula at the mean values for the explanatory variables $(\bar{X}_i)$:

$$\hat{z}_t = \hat{\mu} + \widehat{\emptyset}_1 \hat{Z}_{t-1} + \widehat{\emptyset}_2 \hat{Z}_{t-3} + \widehat{\emptyset}_2 \hat{Z}_{t-12} + \sum_{i=1}^{10} \hat{\beta}_i (\bar{X}_i + S_{i,t}) \tag{5}$$

where $S_{i,t}$ is the shock to significant explanatory variable X_i in time period t.

(3) In the above formula, we introduce a unit change (or unit shock) by increasing the initial value of the shock term $(S_{i,0})$ by one (from zero) for variable X_i, whose marginal effect we are trying to estimate, holding all other variables constant at their respective means. The resulting $\hat{Z}_t$ is recorded.

(4) Similarly, a sustained shock is introduced by increasing all shock values for an explanatory variable X_j by one, holding all other variables constant at their respective means. The $\hat{Z}_t$ calculated based on the sustained shock is recorded.

(5) Marginal effects are calculated as the difference between $\hat{Z}_t$ after the introduction of the shock and $\hat{Z}_t$ in the absence of shock. These are calculated for both scenarios, that is, one scenario where only the initial shock is introduced, and another scenario where the variable of interest suffers a sustained shock.

The estimated marginal effects for the monthly model are illustrated by Figure 5 for the time period of 60 months (i.e., from t = 0 to t = 60). The marginal effects indicate that when temperature rises by 1 °F from the mean, the average hourly load in a specific month increases in the first period, by approximately 14 MW, and then decreases precipitously in the next period by around 6 MW, following a pattern of rise and fall in alternate periods until the marginal effect due to an initial shock to temperature dampens out by the 60th month. A sustained shock to temperature by 1 °F increases the average hourly load in a month such that the new long term mean is also increased to a higher level by approximately 9 MW.

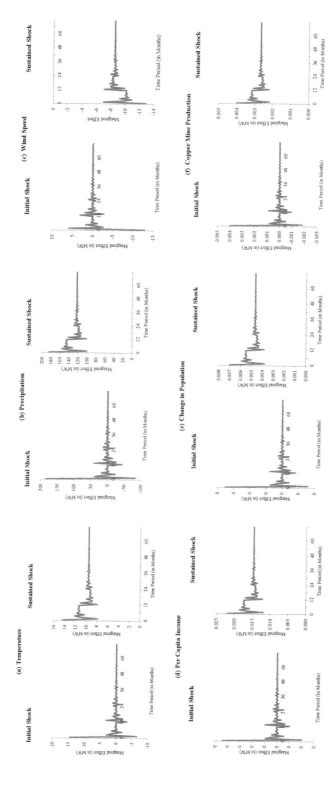

Figure 5. (Color online) Marginal effects for the monthly model: (a) temperature, (b) precipitation, (c) wind speed, (d) per capita income, (e) change in population, (f) copper mine production.

Precipitation, per capita income, population change and copper mine production displays a similar trend. A sustained unit shock has a positive long term impact on average hourly load in a month. However, a unit increase in wind speed brings about a chilling effect, reducing the average hourly load in a specific month by almost 13 MW in the first period. In the next period, this initial shock results in an increase of 6 MW, followed by another crest-trough pattern, but one that is dampening through time. A unit of sustained shock to wind speed negatively impacts the long term value of average hourly load in a month by around 9 MW.

Conclusion

Regional climate models predict hotter and drier weather for the southwestern US, where electricity generation competes for water with large cites and extensive irrigated agriculture. Despite a variety of load forecasting tools available, local electric utilities in this area tend to rely upon past load data to predict future load profiles. We investigate the potential effects of climate change by incorporating weather and socio-demographic variables into a forecasting model for average hourly load by month. We find a 5% improvement in forecast accuracy over a purely autoregressive model.

Policy implications and future research

Improved forecasts of average hourly electricity load by month have the potential to reduce local utilities' electricity costs by minimizing exposure to volatile spot markets when underproduction occurs. Seasonal patterns occur in electricity spot market prices because even modest changes in demand cause significant price spikes when demand is already high (Weron, 2006). More accurate load forecasting models that incorporate climate information may help minimize a utility's need for engaging in the volatile electricity spot markets.

In addition, electric load exhibits strong seasonal fluctuations, with peak demand in the arid deserts of the southwestern US occurring in the summer. Electricity generation is highly dependent on water availability, and water demand is highest in summer months, when water supply is most constrained. An improved understanding of summer peak electricity loads is essential to preparing for the hotter temperatures and more variable precipitation that are expected to accompany climate change. More accurate load forecasts can also aid in reducing the carbon footprint of electric utilities by ensuring efficient use of resources and reduced GHG emissions.

For a more comprehensive picture of the impact of climate change on electricity loads, future work should focus on using dynamically downscaled climate model data to predict future load profiles in light of different climate change scenarios. These predicted load profiles can then inform discussions on the economic and environmental viability of investment in generation technologies. For example, in Arizona, the Renewable Energy Portfolio Standard (RPS) requires electricity providers to supply 15% of their electricity from renewable sources by the year 2025. For this region, solar energy is considered a natural alternative to current generating technologies. However, a barrier to installation of large solar thermal power plants is their high water demand (Pasqualetti & Haag, 2010).

Since 2007, Georgia, Idaho, Arizona, and Montana have each denied permits for conventional power plants because there was not enough water available at the proposed plant site to operate them (Glennon, 2009). Using climate-sensitive load models will facilitate more informed regional planning for electricity generation and water supply, including the water-energy implications of alternative electricity generation technologies.

Acknowledgements

The authors appreciate insightful comments and suggestions provided by Satheesh Aradhyula and Gary Thompson. We acknowledge our research colleagues with the Climate Assessment for the Southwest program at the University of Arizona. This work was supported by that project through the National Oceanic and Atmospheric Administration's Climate Program Office (NOAA), grant NA16GP2578. The statements, findings, conclusions, and recommendations are our own and do not necessarily reflect the views of NOAA, US Department of Commerce, or the US Government Climate Program. We also thank TEP and AZMET for providing the necessary electricity load and weather data.

References

Alfares, H. K. & M. Nazeeruddin. (2002). Electric load forecasting: Literature survey and classification of methods. *International Journal of Systems Science, 33*, 23.

Al-Zayer, J. & Al-Ibrahim, A. A. (1996). Modeling the impact of the temperature on electricity consumption in Eastern province of Saudi Arabia. *Journal of Forecasting, 15*, 97–106.

Amato, A. D., Ruth, M., Kirshen, P., & Horwitz, J. (1996). Regional energy demand response to climate change: Methodology and application to the Commonwealth of Massachusetts. *Climate Change, 71*, 175–201.

Bessec, M. & Fouquau, J. (2008). The non-linear link between electricity consumption and temperature in Europe: A threshold panel approach. *Energy Economics, 30*, 2705–2721.

Chandrasekharan, B. (2011). *Electricity load forecasting improvements as a climate change adaptation* (Unpublished thesis). University of Arizona.

Considine, T. & McLaren, D. (2008). Powering Arizona, choices and trade-offs for electricity policy: A study assessing Arizona's energy future (Draft Report). Los Angeles, CA: The Communications Institute.

Contaxi, E., Delkis, C., Kavatza, S., & Vournas, C. (2006). The effect of humidity in a weather-sensitive peak load forecasting model. *Power Systems Conference and Exposition 2006. IEEE PES*, 1528–1534.

Dominguez, F., Canon, J., & Valdes, J. (2010). IPCC-AR4 Climate simulations for the Southwestern US: The importance of future ENSO projections. *Climatic Change, 99*, 499–514.

Electric Power Research Institute. (2006, December). *Climate change impacts and the electric sector* (Technical Report 1013189). Palo Alto, CA: EPRI.

Engle, R. F., Granger, C. W. J., Rice, J., & Weiss, A. (1986). Semiparametric estimates of the relation between weather and electricity sales. *Journal of the American Statistical Association, 81*, 310–320.

Feinberg, E. (2009, November). *Peak load forecasting*. Advanced Energy Conference, Hauppage, NY.

Garfin, G. (2012). *Southwest climate assessment*. Tucson, AZ: Climate Assessment for the Southwest.

Glennon, R. (2009). *Unquenchable: America's water crisis and what to do about it* (1st ed.). Washington, DC: Island Press.

Henley, A. & Peirson, J. (1997). Non-linearities in electricity demand and temperature: Parametric versus non-parametric methods. *Oxford Bulletin of Economics and Statistics, 59*, 1149–1162.

Hutson, S. S., Barber, N.L., Kenny, J.F., Linsey, K.S., Lumia, D.S., & Maupin, M.A. (2004). Estimated use of water in the United States in 2000. Reston, VA: US Geological Survey.

Hyndman, R. J. & Fan, S. (2010). Density forecasting for long-term peak electricity demand. *IEEE Transactions on Power Systems, 25*, 1142–1153.

Lam, J. C. (1998). Climatic and economic influences on residential electricity consumption. *Energy Conversion and Management, 39*, 623–629.

Mirasgedis, S., Sarafidis, Y., Georgopoulou, E., Lalas, D. P., Mschovitis, M., Karagiannis, F., & Papakonstantinou, D. (2006). Models for mid-term electricity demand forecasting incorporating weather influences. *Energy, 31*, 208–227.

Moral-Carcedo, J. & Vicéns-Otero, J. (2005). Modeling the non-linear response of Spanish electricity demand to temperature variations. *Energy Economics, 27*, 477–494.

Murphy, A. H. (1993). What is a good forecast? An essay on the nature of goodness in weather forecasting. *Weather Forecasting, 8*, 281–293.

Niemuth, N. J. (2004). *Arizona mining update – 2002 and 2003* (Circular No. 108). Phoenix: Arizona Department of Mines and Mineral Resources.

Niemuth, N. J. (2005). *Arizona 2004 mining review*. Phoenix: Arizona Department of Mines and Mineral Resources.

Niemuth, N. J. (2006). *Arizona mining update – 2005* (Circular No. 118). Phoenix: Arizona Department of Mines and Mineral Resources.

Niemuth, N. J. (2007). *Arizona mining update – 2006* (Circular No. 125) Phoenix: Arizona Department of Mines and Mineral Resources.

Niemuth, N. J. (2008). *Arizona mining update – 2007.* (Circular No. 129) Phoenix: Arizona Department of Mines and Mineral Resources.

Pardo, A., Meneu, V., & Valor, E. (2002). Temperature and seasonality influences on Spanish electricity load. *Energy Economics, 24*, 55–70.

Pasqualetti, M. & Haag, S. (2010). A solar economy in the American Southwest: Critical next steps. *Energy Policy, 39*, 887–893.

Psiloglou, B. E., Giannakopoulos, C., Majithia, S., & Petrakis, M. (2009). Factors affecting electricity demand in Athens, Greece and London, UK: A comparative assessment. *Energy, 34*, 1855–1863.

Ranjan, M. & Jain, V. K. (1999). Modeling of electrical energy consumption in Delhi. *Energy, 24*, 351–361.

Sailor, D. J. & Muñoz, J. R. (1997). Sensitivity of electricity and natural gas consumption to climate in the U.S.A. - methodology and results for eight states. *Energy, 22*, 987–998.

Srinivasan, D. & Lee, M. A. (1995, October). Survey of hybrid fuzzy neural approaches to electric load forecasting. In Systems, Man and Cybernetics, 1995. *Intelligent Systems for the 21st Century* (pp. 4004–4008). Taipei, Taiwan: IEEE International Conference.

Suganthi, L. & Samuel, A. A. (2012). Energy models for demand forecasting: A review. *Renewable and Sustainable Energy Reviews, 16*, 1223–1240.

Tanimoto, P. M. (2008). *Forecasting the impact of climate change for electric power management in the Southwest* (Unpublished thesis). University of Arizona.

US Energy Information Administration. (2010). *Electric power annual 2009*. Washington, DC: DOE/EIA–0348.

US Global Change Research Program. (2009). *Global climate change impacts in the United States*. Cambridge, UK: Cambridge University Press.

Weron, R. (2006). *Modeling and forecasting electricity loads and prices: A statistical approach*. New York, NY: John Wiley and Sons.

Willis, H. L. (2002). *Spatial electric load forecasting* (2nd ed.). New York, NY: Marcel Dekker.

Abbreviations

ACF	Autocorrelation function
AR	Autoregressive
ARIMA	Autoregressive Integrated Moving Average
ARMA	Autoregressive Moving Average
AZDMMR	Arizona Department of Mines and Mineral Resources
AZMET	Arizona Meteorological Network
GCM	Global Climate Model
GHG	Green House Gas
FGLS	Feasible Generalized Least Squares
IEA	International Energy Agency
LSTR	Logistic Smooth Threshold Regression

LTLF   Long-term load forecasting
MAPE   Mean Absolute Percentage Error
MISS   More Intense Summer Shift
MSA   Metropolitan Statistical Area
MTLF   Medium-term load forecasting
MW   Mega-Watt
OLS   Ordinary Least Squares
RMSE   Root Mean Squared Error
RPS   Renewable Energy Portfolio Standard
STLF   Short-term load forecasting
TEP   Tucson Electric Power
USGCRP   US Global Change Research Program
USGS   US Geological Survey

Appendix 1. Calculating annual cost savings from monthly model 2009 forecasts

To calculate annual cost savings from the use of improved forecasts, we follow the steps listed below:

(1) We obtain estimates of average per-unit generation costs for different technologies. In our study, we used 2009 estimates provided by Timothy Considine in his study 'Powering Arizona' (2008): \$51.5/MW for scrubbed coal, and \$106/W for conventional gas.
(2) We calculate forecast errors for each month of the year 2009, whose absolute value is multiplied by the number of hours in a month, per unit generation cost, and a spot market indicator to obtain losses due to forecast errors for the respective model. We apply this calculation to both the Full and Basic model forecasts for the year 2009. The spot market indicator is defined in the following manner:

$$Spot\ Market\ Indicatior = \begin{cases} 1\ if\ forecast\ error \leq 0 \\ 2\ if\ forecast\ error > 0 \end{cases}$$

where $Forecast\ Error = Actual\ Load - Forecasted\ Load$.

Therefore a positive forecast error implies under-production by the electric utility, necessitating its entry into the spot market (spot market indicator = 2), where it pays twice as much for a unit of electricity than if it generated it (i.e., Case B). Under this scenario we assume that losses in the spot market are asymmetrical, depending upon whether the utility is under-producing or over-producing electricity. We incorporate this asymmetry by assuming that the loss per MW incurred by the electric utility in case of under-production is twice the generation cost per unit of electricity

However, in absence of specific data regarding the spot market, we also assume a simple scenario where the losses per MW incurred by the electric utility in case of over-production or under-production are equivalent (i.e., Case A). This means that even if the utility generates less electricity than required, it incurs a purchase price per unit of electricity in the spot market that we assume is equal to the cost of generation per unit.

(3) Next, we calculate annual cost savings using the following formula:

Annual Cost Savings (in\$)	=	Total losses due to forecast errors from Basic Model	−	Total losses due to forecast errors in Full Model

The joint impact of drought conditions and media coverage on the Colorado rafting industry

Karina Schoengold[a], Prabhakar Shrestha[b] and Mark Eiswerth[c]

[a]Department of Agricultural Economics and School of Natural Resources, University of Nebraska; [b]School of Natural Resources, Rm 249 (#26) University of Nebraska; [c]Department of Economics, University of Northern Colorado

Decreased instream flows are thought to negatively impact river recreation, such as whitewater rafting. Runoff declines are often accompanied by high temperatures, wildfires, and associated media coverage. Very few studies have examined the impact of these accompanying factors, and none have disentangled their influence. Using regression analysis of data from the Arkansas River in Colorado, we find that reduced flows usually decrease the number of whitewater rafting customers, although very high flows also deter rafting activities. More importantly, negative media coverage of wildfires also appears to have adverse impacts on rafting tourism, controlling for instream flows and weather.

Introduction

Changes in the frequency of extreme weather events are among the most serious challenges to society in coping with a changing climate (Karl et al., 2008). Based on observed increases in global average air and ocean temperatures, widespread melting of snow and ice, and rising global average sea level, the Intergovernmental Panel on Climate Change (IPCC) in its Fourth Assessment Report (FAR) concludes that warming of the climate system is unequivocal. In the Working Group II Report on Impacts, Adaptation and Vulnerability, the IPCC projects that warming in the western mountains of the North American region will cause decreased snowpack and reduced summer flows, exacerbating competition for over-allocated water resources (Field et al., 2007). In addition, dry conditions and hot summer weather are known to increase the frequency and severity of wildfires. The drought-stricken summer of 2012 saw several wildfires in Colorado, including the destructive Waldo Canyon fire near Colorado Springs, which destroyed about 350 homes (Gorski, 2012).

While it is still too early to measure the full economic impact of the drought and wildfires of 2012, results of similar conditions in Colorado in 2002 (particularly the Hayman fire) can provide useful evidence about the impact of drought and wildfires on tourism. The Hayman fire started on 8 June 2002, and lasted for approximately 20 days. It covered over 138,000 acres and destroyed 132 homes (Graham, 2003). Previous studies have found a large negative economic impact of the Hayman fire on Colorado tourism, much of which is attributed to poor communication about the fire's extent. In particular, in June 2002, following an aerial tour of wildfires occurring in Colorado, Governor Owens commented that 'all of Colorado is burning' (Page 12, New York Times, 2002).

Such comments and widespread media coverage are thought to have caused an immediate impact on the tourism industry. For example, the number of trips to the state of Colorado dropped off sharply in 2002, during the July to October period (State of Colorado, 2002). This decline was proportionally larger in Colorado than in other states. The number of domestic travelers to Colorado by air transportation in 2002 was 5.1 million, the lowest in the 1996 to 2011 period (Dean Runyan & Associates, 2004; Dean Runyan & Associates, 2012). The number of whitewater rafting customers plummeted in 2002 relative to the previous year (Figure 1 shows customer numbers by year for Colorado's Arkansas River, as an example). This occurred despite the fact that the Hayman fire was not very close to the Arkansas River and was not an impediment to rafting. With hindsight as an ally, officials took a more cautious approach during the 2012 Waldo Canyon fire, with tourism officials attempting to educate the public on the fire's relatively small impact on recreational opportunities (Verlee, 2012).

The focus of our study is to measure the effects of weather conditions, associated media reports, and economic conditions during the 2002 season on one segment of Colorado's tourism industry in one geographic location; specifically, white water rafting on the Upper Arkansas River. By using multiple years of daily visitation data, we are able to measure the impact of instream flow and economic variables on the number of rafting customers. We find evidence that the number of rafting customers was expected to decrease due to low instream flow, but that media reports may have exacerbated that impact.

Background

Colorado is a semiarid state where drought is a frequent phenomenon. Despite limited water resources, Colorado's population is growing rapidly, with increases of 31% and 17% in the 1990s and 2000s, respectively (US Census Bureau, 2001). Several studies

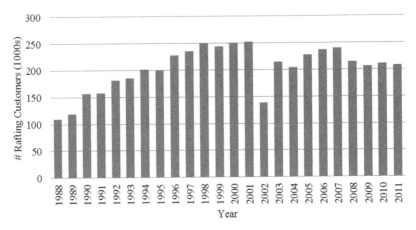

Figure 1.   Total number of rafting customers per year on the Arkansas River (1988 – 2011).
Source: Colorado River Outfitters Association (2011).

(e.g., McKee et al., 2000; Nichols, Murphy, & Kenney, 2001) show Colorado's growing vulnerability to water shortages and drought.

A recent report published by the Colorado Water Conservation Board (CWCB) suggests that Colorado's annual mean temperature has increased about 2°F in the past 30 years. Computer models project that Colorado's temperature will warm 2.5°F by 2025 and 4°F by 2050 relative to the 1950 to 1999 baseline (CWCB, 2008). This implies Colorado's summer monthly temperatures will be as warm as, or warmer than, the hottest 10% of summers in the 1950 to 1999 period. Hydrologic projections also show a decline in runoff for most of Colorado's river basins by the mid twenty-first century (CWCB, 2008). Changes in instream flow intensity and timing will directly affect whitewater rafting in the future (CWCB, 2008).

The summer of 2002 was preceded by dry and warm winters in 2001 and 2002. Statewide precipitation in February 2002 was only 56% of average (Natural Resources Conservation Service [NRCS], 2002a). By 31 May 2002, snowpack in the South Platte, Upper Arkansas and Upper Colorado was at 23%, 34% and 28%, respectively, of the long-term average (NRCS, 2002b). Similarly, water storage in these basins dropped to just over 80% of the long-term average. By summer, the entire state of Colorado was in an extreme drought (National Oceanic and Atmospheric Administration [NOAA], 2002), which increased stress on instream flow users.

Two sectors of the Colorado tourism industry that are most dependent on precipitation are skiing and whitewater rafting. Skiing relies on precipitation in the form of snow, while rafting relies on precipitation's effects on instream flow. Commercial rafting activity in Colorado generates $54 million in expenditures; the economic contribution of each acre-foot (AF) of water ranges from as high as $352/AF on the Arkansas River to as low as $18/AF on the Colorado River through Glenwood Canyon (Loomis, 2008). This particular measure of the value of a non-consumptive use (i.e., state income generated by rafting per acre-foot of instream flow) for the Arkansas River is competitive with values generated by irrigated agricultural crops such as alfalfa and corn.

Other studies show the direct, indirect and induced impacts of rafting activity on regional economies associated with rafting sites (Bowker, English, & Donovan, 1996; Johnson & Moore, 1993; Leones, Colby, Cory, & Ryan, 1997). Similarly, other studies (e.g., Cordell, Bergstrom, Ashley, & Karish, 1990) show regional economic impacts of a broader suite of water-based river recreation activities (e.g., rafting, swimming and tubing). Several newspaper articles published during the 2002 drought mention negative impacts to the rafting industry (e.g., *Pueblo Chieftain*, 23 August 2002 [Harmon, 2002]; *The Gazette (Colorado Springs)*, 27 July 2002 [Darling, 2002]; *Denver Post*, 28 July 2002 [Blevins, 2002]). However, a comprehensive and detailed investigation into the rafting outfitters' impacts, challenges, operational changes and mitigation strategies is missing, to date.

Colorado has 27 rivers where commercial whitewater rafting takes place. Table 1 shows visitation numbers for the six most-popular rivers and the entire state for the period 2000 to 2007. Total number of rafting customers decreased substantially in the 2002 season relative to the previous two years. Only in 2007, five years after 2002, did the number of rafting customers surpass 2001 levels. Among all rivers in Colorado, the Arkansas River serves the highest number of rafting customers (almost 50% of Colorado's rafting customers). In fact, the Arkansas River is one of the most rafted rivers in the world.

Data and summary statistics

We combine several datasets for our econometric analysis of rafting demand. The first is a firm-level daily customer record for the period 2000 to 2006, obtained from the Arkansas

Table 1.   Number of commercial rafting customers in Colorado (2000-2007), by river.

River	2000	2001	2002	2003	2004	2005	2006	2007
Animas	29,000	42,000	12,000	34,500	35,470	52,700	42,500	44,322
Arkansas	250,861	252,213	139,178	214,555	203,840	228,091	237,160	239,887
Clear Creek	13,616	20,798	7,498	24,495	20,115	32,357	36,889	49,190
Colorado– Glenwood	57,265	55,829	42,581	56,876	58,751	57,712	62,652	65,502
Colorado – Upper	43,579	34,373	36,365	41,054	35,500	33,603	31,073	31,997
Poudre	29,012	34,192	26,004	34,164	31,042	36,088	34,533	37,824
All rivers in Colorado	503,524	515,704	309,129	462,884	445,816	510,544	510,304	539,222

Source: Colorado River Outfitters Association (2008).

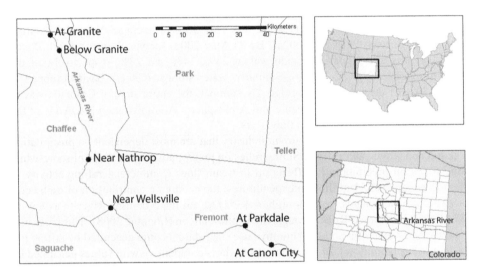

USGS Gauge Stations along Arkansas River

Figure 2.   (Color online) Map of Arkansas River Basin, Colorado, including locations of the six USGS gauging stations used in the analysis.
Source: USGS.

Headwaters Recreational Area (AHRA). Approximately 55 firms operate on the river. The company-level data was aggregated to determine the total number of customers on the river each day. Figure 2 shows a map of our study area.

The second dataset includes daily average instream flow and daily maximum air temperature, obtained from the United States Geological Survey (USGS). The USGS has numerous water gauge stations at different sites along the river. However, these sites have not been in continuous existence; some have been closed while others have been recently initiated. We use six different USGS gauge stations that have complete records of daily instream flow and maximum daily temperature for the period 2000 to 2006. The six sites are located across different sections of the Arkansas River.

We also include two economic variables that potentially affect overall demand for recreation: (1) national quarterly GDP values from the Bureau of Economic Analysis (measured in 2005 dollars and available at: http://www.bea.gov/national/index.htm), and (2) weekly regular conventional gasoline prices from the US Energy Information Administration (adjusted to constant 2000 dollars using the CPI, and available at: http://www.eia.gov/ oil_gas/petroleum/data_publications/wrgp/mogas_history.html).

To determine temperature and streamflow for each section of the river, we use the gauge that is closest to the section. Values for each section are weighted by the proportion of customers on that section, and summed, to create a weighted-average temperature and streamflow observation for the entire study area for each day. The total number of customers who raft during different months of the period 2000 to 2006 is shown in Figure 3. Very few customers raft in April and September, so these months are excluded from the empirical analysis. May through August is the primary rafting season; the number of customers peaks in July and declines through the rest of the season. The mean, standard deviation, minimum and maximum values of the variables of interest from the aggregated dataset are shown in Table 2. The aggregated dataset contains one observation of the dependent variable (number of commercial rafting customers) for each day on which some rafting activity occurred.

Figures 4 and 5 show scatterplots of the daily number of customers versus daily average instream flow, organized by month. Figure 4 includes only the 2002 season, while Figure 5 covers the period 2000 to 2006 excluding data from 2002. We present these in two diagrams because instream flow levels in 2002 are significantly lower than in other years. Comparison of Figures 4 and 5 shows that the general distribution of customers

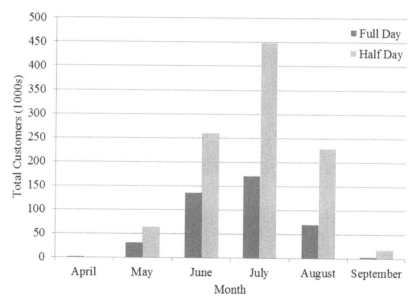

Figure 3.   Total number of customers by month and type of trip for 2000–2006.

Table 2.   Summary statistics for regression variables.

Variable	Mean	Std. Deviation	Minimum	Maximum
Customers per day	1636.1	1184.6	2	5154
Instream flow (daily avg; cfs)	893.9	604.2	179.4	3403.6
Air temperature (daily max; C)	17.57	2.70	5.80	24.90
Gas price (weekly; $ per gallon)	1.72	0.38	1.31	2.60
Quarterly GDP (billion $)	11972.0	605.5	11248.8	12950.4

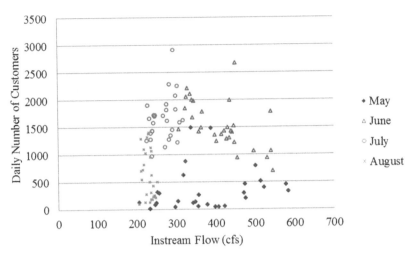

Figure 4.   Scatterplot of daily number of customers versus daily instream flow during 2002.

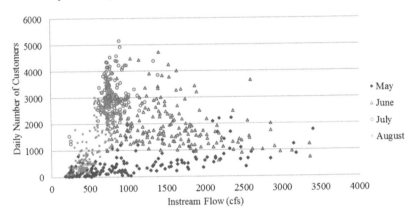

Figure 5.   Scatterplot of daily customers versus daily instream flow during 2000–2001 and 2003–2006 (i.e., 2002 data deleted).

across months in 2002 is similar to that in other years, with the peak number of customers occurring in medium flow conditions, and in June and July.

In Figure 4, the number of customers is highest when flows are between 250 and 450 cubic feet per second (cfs), but rarely exceeds 2000 customers per day. In comparison, in Figure 5, the daily number of customers frequently exceeds 2000 customers per day. In addition, instream flow for 'normal' years (2000 to 2006, excluding 2002) is several times higher than in 2002; the majority of customers during normal years visit when instream flow is between 700 and 1700 cfs. Instream flow never exceeded 600 cfs in 2002 (Figure 4).

Econometric model

We estimate an econometric model to test the general impact of weather variables (instream flow and air temperature) on the number of commercial rafting customers on the Arkansas River, and to examine how drought conditions and other factors in 2002 affected the rafting industry. Most rafting customers make advanced reservations for their trips, at which time air temperature and instream flow for the scheduled day of the trip are uncertain. Thus, we model demand as a function of weather variables leading up to the scheduled trip, keeping in mind customers may cancel reservations if they anticipate unfavorable conditions. The flow variable we use is 'average flow for the week before a completed trip', and our temperature variable is 'maximum temperature for the day before a completed trip'.

We estimate the following model using daily data:

$$NC_t = \beta_0 + \beta_1 F_t + \beta_2 F_t^2 + \beta_3 T_t + \beta_4 Y02 + \beta_5 GDP_t + \beta_6 Gas_t + \beta_7 Weekend_t$$

$$+ \sum_{m=1}^{3} (\gamma_{1m} month_m + \gamma_{2m} month_m * Y02) + \epsilon_t,$$

where $\epsilon_t = \rho \, \epsilon_{t-1} + \mu_t$ is the error term accounting for serial correlation; NC_t is number of customers on day t; F is average instream flow for the previous week (cfs); T is maximum daily temperature for the previous day; $Y02$ is a binary (0-1) indicator variable for the 2002 season; GDP is the quarterly GDP value; Gas is the weekly average price of regular gasoline; $month_m$ is a vector of binary (0-1) indicator variables (where $m =$ June, July, and August; May is the omitted month); and $weekend$ is a binary variable that equals one if the day is a Saturday or Sunday. The average number of daily rafting customers is 2023 for weekend days and 1482 for weekdays. Customers visiting from outside the state of Colorado may raft any day of the week, but in-state visitors are more likely to take rafting trips on weekends. The parameters β, γ, and ρ are to be estimated in our analysis.

We expect the relationship between instream flow and number of customers to be quadratic. As instream flow increases over a normal range, it is more fun to be in the river and thus more appealing to both tourists and avid rafters. However, when instream flow is very high, it might trigger flood warnings and make rafting dangerous. Temperature is expected to have a positive relationship with the number of customers. More people are expected to want to be on the river and raft as maximum daily air temperature increases. If the previous day is warm, more people are likely to book rafting trips, whereas they may cancel if the day before a scheduled trip is cool.

Binary (0-1) indicator variables are included to incorporate the effects of the 2002 year, as well as variation in visitation during different months of the season. The $Y02$ variable is meant to capture the change in customer numbers in 2002 due to wildfires in Colorado that year and perhaps reduced travel in the aftermath of the 11 September attacks in 2001.

The *month* indicator variables reflect that visitation during a rafting season is not distributed evenly across all months. We include separate interaction terms between each month and 2002 (i.e., $month_m * Y02$) because several outfitters suggested that the decrease in rafting customers that year was not solely due to drought and associated low flow conditions. They hypothesized that media coverage of wildfires during the season affected the number of customers, despite an absence of wildfires in the Arkansas River Basin. Month-specific interaction terms allow us to see if this hypothesized effect can be measured in the data.

Lastly, economic variables are captured through the GDP variable, which provides information about the general state of the economy, and the gas price variable, which measures a portion of travel cost for rafters. We expect rafting demand to be positively related to GDP and negatively related to gas price.

We estimate three regression models. The first, 'basic model', includes the following regressors: instream flow, instream flow squared, maximum daily temperature, and the

2002 indicator variable. The second model, 'month effects', includes all of these variables plus monthly binary indicator variables. The third, 'full model', includes all of the variables in the second model, plus interaction terms between monthly indicator variables and the 2002 indicator variable.

Because we are using time series data, we test each model for serial correlation. We find evidence of serial correlation in all three regressions; the Durbin-Watson test statistic is 0.522, 0.712, and 0.767, respectively. The correction procedure in Stata accounts for gaps in the data (i.e., days on which zero rafting trips occurred), and recalculates the serial correlation value for each period.

Results

Results from our three models are presented in Table 3. As expected, there exists a quadratic relationship between number of customers and instream flow. As the flow-level increases, the number of customers increases up to a certain point; however, when the flow-level exceeds this point, the number of customers declines. Previous literature suggests that demand for rafting is positively correlated with instream flow or water in storage (Loomis, 2008; Walsh, Ericson, Arosteguy, & Hansen, 1980). Our results show that this is true over a range of instream flow levels, but when instream flow is very high, demand decreases for rafting because it is considered more dangerous in fast-moving water conditions. The inflection point for this quadratic relationship, in our data and estimation results, is at 1758, 1649, and 1673 cfs, respectively, for models (1), (2), and (3). According to Figure 5, these values are within the range of normal flow conditions, but are well above the average flow of 893.9 cfs.

The relationship between maximum daily temperature and number of customers is positive and statistically significant, as expected. Coefficients on economic variables (gas price and GDP) also have the expected signs, but are not significant in any of the regression results.

The coefficient on the 2002 binary indicator variable is not significant in any of our regressions. However, in model (3), interaction terms between monthly indicator variables and the 2002 indicator variable are significant, particularly for July and August. This suggests that, after controlling for flow, temperature and economic variables, the constant term for 2002 does not differ significantly from that in all other years. However, the effect of July and August on rafting demand is significantly different for 2002 than for all other years.

Model (1) measures the average effect of the 2002 year, after controlling for instream flow, air temperature, weekend days, and economic variables. Instream flow was significantly lower in 2002 than in other years (average daily value of 321.8 cfs and 989.2 cfs, respectively). Thus, coefficients on flow and flow-squared imply an average reduction of 941.6 customers per day, or a little over 50% of average visitation of 1737.3 during non-drought years. Figure 6 shows actual versus predicted daily customer numbers based on the results of model (1); actual and predicted daily customers are averaged across seven-day increments, so they can be reported by week. For years other than 2002, actual and predicted values are also averaged over the six years. We use weekly averages to help illustrate changes throughout the season because daily values are highly variable. Figure 6 clearly shows that not including monthly indicator variables (i.e., binary variables 'June', 'July', and 'Aug') underestimates visitation in the peak part of the season (weeks 8–15). It also shows that this difference is less pronounced in 2002. To correct for this, we move on to results for model (2).

Table 3.   Regression results; dependent variable is 'number of customers per day'.

Variable	Model (1): Basic	Model (2): Month Effects	Model (3): Full
Constant	−3658.54	−2899.73	−3003.90 *
	(−1.01)	(−1.60)	(−1.76)
Lagged Flow (cfs)	2.25 ***	2.77 ***	2.71 ***
	(4.09)	(7.67)	(7.81)
(Lagged Flow)^2	−0.00064 ***	−0.00084 ***	−0.00081 ***
	(−3.60)	(−6.97)	(−6.96)
Previous Day Temperature (C)	55.69 ***	18.29 ***	59.73 ***
	(3.06)	(3.02)	(3.26)
Weekend	546.80 ***	536.20 ***	533.39 ***
	(15.74)	(14.83)	(14.63)
Y02	−201.23	−64.09	309.95
	(−0.51)	(−0.33)	(1.10)
June		691.73 ***	703.08 ***
		(4.99)	(4.81)
July		1517.09 ***	1652.43 ***
		(9.96)	(10.69)
August		712.23 ***	794.21 ***
		(4.51)	(4.97)
June*Y02			−85.34
			(−0.25)
July*Y02			−741.42 **
			(−2.08)
Aug*Y02			−645.84 *
			(−1.79)
Gas Price	−506.97	−275.22	−266.37
	(−0.87)	(−0.93)	(−0.96)
Quarterly GDP	0.32	0.14	0.13
	(0.87)	(0.76)	(0.79)
R^2	0.238	0.383	0.406
Rho	0.847	0.675	0.652
Durbin−Watson (pre)	0.522	0.712	0.767
Durbin-Watson (post)	2.387	2.191	2.170
Sample Size (N)	854	854	854

Note: T-statistics are shown below the coefficients. Statistical significance is indicated at the 1% (***), 5% (**), and 10% (*) levels. All results are corrected for AR(1) serial correlation.

Model (2) includes indicator variables for the months of June, July and August (the omitted month is May), in addition to regressors present in model (1). When these indicator variables are introduced, number of customers is found to increase as the season progresses, peaking in July and then decreasing thereafter (Table 3 and Figure 7). Each of the three monthly indicator variables is statistically significant at the 1% level. The general trend of more customers during June, July, and August is not surprising because summer is the prime season for family vacations. Figure 7 shows actual versus predicted values from model (2); it reports weekly averages in the same manner as in Figure 6.

A comparison of Figures 6 and 7 shows that inclusion of monthly indicator variables improves model (2)'s predictive power throughout the season, with predicted customer numbers close to actual customer numbers in most weeks. However, Figure 7 still shows a discrepancy between predicted and actual values in 2002. In particular, actual customer numbers are higher than the model predicts for the early season (weeks 1–9), but lower than

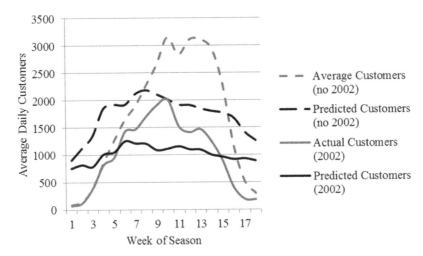

Figure 6. Comparison of actual versus predicted customers, model (1).

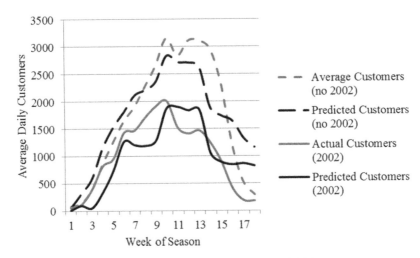

Figure 7. Comparison of actual versus predicted customers, model (2).

the model predicts for the mid-to-late season (weeks 10–13 and 16–18). The season starts in May, so this indicates an unexplained shift in demand for rafting in mid-to-late June. Model (2) performs better for the other years, although actual customer numbers do show a more dramatic drop than predicted between the beginning and end of August, possibly due to children going back to school and thus fewer family vacations.

Model (3) includes both monthly indicator variables as well as interaction terms between these monthly indicators and the 2002 indicator variable (Table 3; Figure 8). While not statistically significant, it is interesting that the 2002 indicator variable's coefficient is negative in models (1) and (2), but positive in model (3): -201.23, -64.09, and 309.95, respectively. This suggests the actual number of customers in the base month of 2002 (May) was higher than predicted by climate and economic variables alone. The coefficient on the interaction term between the June and 2002 indicator variables is also insignificant. These combined results suggest that, after accounting for temperature, instream flow and

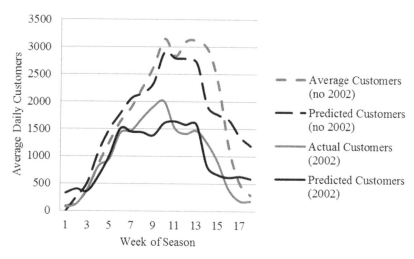

Figure 8.   Comparison of actual versus predicted customers, model (3).

economic variables, the first part of the 2002 season (May and June) was not significantly different than in other years.

As the 2002 season progressed, however, the actual number of customers declined by a larger proportion than in other years. The negative and statistically significant coefficients on the July and August interaction terms with the 2002 indicator are evidence of this. Figure 8 shows that inclusion of seasonal effects specific to 2002 in model (3) results in predicted customer numbers that are much closer to actual visitation, compared to Figure 7 for model (2). In summary, results of the empirical analysis show that reduced instream flow and economic activity partially explain the decline in customer numbers during the 2002 season, but are not sufficient to explain the more dramatic reduction in the second-half of the season, relative to other years.

Although individual rafters' responses are obscured by looking at industry-level data, this aggregate analysis is necessary to understand overall impact. An overall decline of 40 to 50% in the number of customers was observed in the year 2002. However, econometric model (3) suggests that the actual number of customers in 2002 was equal to or higher than model (2) predicts for the first two months of the season, while lower than model (2) predicts for later months (July and August). This implies something occurred during June that caused a larger than usual decline in subsequent rafting demand. The decline in rafting numbers after June cannot be attributed solely to lower instream flow or lower economic activity. In the following section we highlight results of interviews with rafting outfitters that provide insight into potential reasons for the unusually large decline.

Interviews with outfitters

One of the primary goals of this study is to measure and understand the effects of the 2002 drought, wildfire, and related impacts on the rafting industry. Doing so requires combining the quantitative results of regression analysis with a qualitative understanding of rafting outfitters' responses to and perceptions of the season. In addition to our quantitative analysis of visitation numbers, we also conducted interviews with several outfitters that have commercial rafting operations on the Arkansas River.

Outfitters used a variety of strategies to cope with low instream flows in 2002. Some suspended their business earlier in the season than usual. Others continued to operate in an attempt to recover some of their investments. Some saw low instream flows as an opportunity to provide rafting trips on sections of the river that are usually too difficult for most customers.

Most outfitters expressed serious concern about the effects of the 2002 season on their business. They all agreed there were multiple and complex problems during the year, including drought, low instream flow, wildfires, and a poor economy after the 11 September attacks. All of the outfitters experienced a significant number of reservation cancellations after Governor Owens' comment about the wildfires in Colorado (i.e., 'All of Colorado is burning'). Some of them believed the media's portrayal of Colorado during summer 2002, as a dangerous destination for visits and outdoor fun, had large impacts on their rafting numbers.

Detailed visitation information is not available for many other rivers, but we try to determine if the decline in 2002 was unique to Colorado or pervasive across other rivers in the US. Figure 9 shows total customer numbers across seven major rivers in the US for the period 1995 to 2004. A close look at the year 2002 shows there were two rivers among the seven that observed a decline in customer numbers greater than 20%: the Poudre and the Arkansas, both in Colorado. The other rivers did not see declines of that magnitude.

We used LexisNexis to search for any reports of adverse effects of drought in 2002 on the whitewater rafting industry in Tennessee, California, or West Virginia. We found no evidence of such an effect in Tennessee or California. An article from West Virginia actually highlighted the benefits to that state's rafting industry of the 2002 drought in the western US (McCoy, 2002). This anecdotal evidence suggests a cause that was local to Colorado, or at least regional to the Mountain West, and not pervasive across the nation.

One of the major events of summer 2002 was the Hayman fire, which affected the Front Range of Colorado. Wilhelmi, Thomas, & Hayes (2005) interviewed several tourism operators, state and county officials, and university extension agents who identified wildfire as a major factor in the overall summer tourism decline. Direct impacts of the wildfires included national and state park closures, bans on campground fires, health threats due to

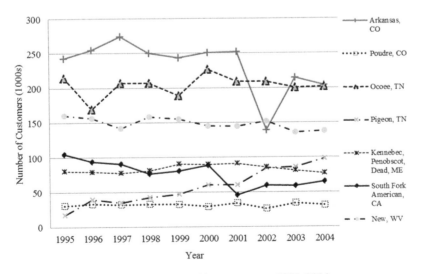

Figure 9.   National summary of whitewater rafting customers, 1995–2004.

Source: Personal communication with D. Brown of American Outdoors Association on February 2 and February 9, 2007.

flames and smoke, and nationwide media coverage of wildfires occurring in Colorado. Park closures did not occur in the immediate vicinity of the Arkansas River, but it seems likely that media coverage had an effect well beyond the Hayman fire's actual perimeter.

Conclusions

Our results indicate that demand for rafting increases as instream flows increase. This finding is consistent with previous studies (e.g., Loomis 2008; Walsh et al., 1980). Our results are somewhat different, however, in that we find a quadratic relationship between instream flow and number of customers. Most of the time, the Arkansas River's actual instream flow is in the 'increasing portion' of the commercial-rafting demand curve.

Our results also indicate that the decline in whitewater rafting customer numbers on Colorado's Upper Arkansas River in 2002 was associated not only with drought-related reductions in instream flow and economic variables, but also with other events occurring in Colorado at that time. Specifically, interviews with rafting outfitters point us toward potential effects of the media message that 'all of Colorado' was on fire, relayed on national television in June 2002. Immediately following these media reports, outfitters experienced cancellations of many summer reservations. Because the industry only operates four months of the year, and relies heavily on advanced reservations, these cancellations had a large impact on their economic activity. This highlights the tourism industry's sensitivity to unfavorable media coverage of catastrophic natural events.

One of the more interesting results from this work is a comparison of predicted and actual customer numbers. Specifically, after controlling for variation in economic and weather variables, the actual number of rafters in May and June of 2002 was similar to predictions for other years. However, the actual number of rafters was lower than predicted for other years (controlling for economic and weather variables) from July through the end of the 2002 rafting season. In other words, our econometric analysis appears to confirm rafting outfitters' stated belief that public comments and subsequent media coverage was the main factor behind business declines during 2002.

As stakeholders seek possible policy interventions for addressing unintended negative economic consequences of media coverage, we suggest more research on the link between negative media portrayal and actual tourist behavior. Very little research has focused on this topic to date. Lexow and Edelheim (2004) conclude that mass media plays an important role in tourists' decisions. Our research provides empirical evidence to support the claim that there was some impact of mass media's coverage of wildfires in Colorado, or public comments about those wildfires, on rafting customer numbers in summer 2002.

Policy implications

We offer three policy suggestions for lessening the producer and consumer impacts of events that do not directly affect the quality of whitewater rafting or other environmental service flows (e.g., a wildfire located far away from a rafting site). Our suggestions are not limited to Colorado; they may be applicable to other recreation-dependent economies.

First, in states where rafting tourism is important, a position could be created or strengthened for an individual to serve as a Media Relations Officer for the whitewater rafting industry. This person could serve as a liaison between the state's Department of Tourism, the Governor's Office, and the tourism industry to coordinate information that might affect the whitewater rafting industry before it is distributed to the media. In Colorado, for example, the Colorado River Outfitters Association's Executive Director now works to coordinate with the Colorado Tourism Office, the lodging industry, and other entities and stakeholders. This type of effort improves the consistency and accuracy of information, and counters incorrect information or negative impressions created by fragmented reporting. Additionally, such liaisons can work to establish marketing campaigns to attract customers when upcoming seasons are expected to involve lower than average flows. Particular segments of the customer base may actually prefer lower flows, for example families or large parties.

Second, we suggest rafting outfitters continue to explore the potential benefits from diversification, both across geographic space as well as recreational activities. Many outfitters operate solely in a few sections of the river, which may limit their ability to attract a broad set of customers. For example, an outfitter working on a technical section of the river with difficult whitewater rapids may receive only avid or frequent rafters. By diversifying and operating in lower levels of the river, an outfitter could broaden their customer base to include families vacationing in Colorado during the summer. To diversify could also mean expanding the business from pure rafting to other aspects of tourism like jeeping, biking, fishing, and hiking. Some outfitters in Colorado have explored such diversification (e.g., fly fishing, a zipline attraction, etc.), but we believe further similar efforts could prove beneficial. Many people who come to the Mountain West are receptive to a broad array of outdoor recreation activities.

Our third suggestion relates specifically to the linkage between wildfires and decisions of recreators and tourists. Wildfires have long been a fact of life in the arid and semiarid West and will continue to be so. At the same time, capabilities in information sharing (particularly web-based) have evolved rapidly in recent years and may offer opportunities to the rafting industry that did not exist several years ago. In the same way that websites currently offer information on wildfires (e.g., http://www.inciweb.org), the whitewater rafting industry could expand web-based dissemination of real-time information on how various wildfires are affecting the quality of whitewater rafting attributes, such as access, water quality or air quality, on a particular rafting section. Real-time transparent reporting of information may result in more cancellations during times when there are *bona fide* negative impacts from a wildfire, but would likely increase customer confidence over the long run; help reassure long-distance tourists during times when there are wildfires but no impacts on rafting sites; and more efficiently distribute recreation and tourism across time and space.

Acknowledgements

The authors would like to thank the National Oceanic and Atmospheric Administration and the USDA Risk Management Agency for financial support of this research. We would also like to thank Joe Atwood, Christopher Goemans, Michael Hayes, Charles Howe, and Raymond Supalla for useful suggestions and feedback on this research. Several members of the whitewater rafting community were instrumental in collecting the data and developing an understanding of the impact of drought: Christina Alvord (Western Water Assessment), John Kreski (Arkansas Headwaters Recreational Area), Joe Greiner (Colorado River Outfitters Association) and Greg Felt. Editors Dannele Peck, Jeffrey Peterson, and two anonymous referees greatly improved the manuscript. Of course, we take responsibility for all remaining errors.

References

Blevins, J. (2002, July 28). Drought, fire, economy pummel tourism: Businesses struggle amid State's woes. *The Denver Post*, p. K1.

Bowker, J. M., English, D. B. K., & Donovan, J. A. (1996). Toward a value for guided rafting on southern rivers. *Journal of Agricultural and Applied Economics, 28*, 423–432.

Colorado River Outfitters Association. (2008). *Commercial river use in the State of Colorado 1988–2008*. Retrieved from http://croa.org/pdf/2006_Commercial_Rafting_Use_Report.pdf

Colorado River Outfitters Association. (2011). *Commercial river use in the State of Colorado 1988–2011*. Retrieved from http://croa.org/media/documents/pdf/2011-commercial-rafting-use-report-final.pdf

Colorado Water Conservation Board. (2008). *Climate change in Colorado: A synthesis to support water resources management and adaptation*. Retrieved from http://cwcb.state.co.us/NR/rdonlyres/B37476F5-BE76-4E99-AB01-D37E352D09E/0/ClimateChange_FULL_Web.pdf

Cordell, H. K., Bergstrom, J. C., Ashley, G. A., & Karish, J. (1990). Economic effects of river recreation on local economies. *Water Resources Bulletin, 26*, 53–60.

Darling, D. (2002, July 27). Drought, declining tourism bring down numbers of rafters on Arkansas River. *The Gazette* (Colorado Springs). Retrieved from Lexis-Nexis.

Dean Runyan & Associates. (2004). *The economic impact of travel on Colorado 1996–2003*. Study commissioned by the Colorado Tourism Office.

Dean Runyan & Associates. (2012). *The economic impact of travel on Colorado 1996–2011*. Study commissioned by the Colorado Tourism Office.

Field, C. B., Mortsch, L. D., Brklacich, M., Forbes, D. L., Kovacs, P., Patz, J. A., Running, S. W., & Scott, M. J. (2007). North America. In M. L. Parry, O. F. Canziani, J. P. Palutikof, P. J. van der Linden, & C. E. Hanson (Eds.), *Climate change 2007: Impacts, adaptation and vulnerability. Contribution of Working Group II to the fourth assessment report of the Intergovernmental Panel on Climate Change* (pp. 617–652). Cambridge: Cambridge University Press.

Gorski, E. (2012, October 14). Neighborhood devastated by Waldo Canyon fire rising again. *The Denver Post*. p. A1.

Graham, R. T. (2003). *Hayman fire case study* (General Technical Report RMRSGTR-114). Ogden, UT: US Department of Agriculture, Forest Service, Rocky Mountain Research Station.

Harmon, T. (2002, August 23). Drought, fear of fires hurt Colorado River rafting companies. *The Pueblo Chieftain*. Retrived from Lexis-Nexis.

Johnson, R. L. & Moore, E. (1993). Tourism impact estimation. *Annals of Tourism Research, 20*, 279–288.

Karl, T. R., Meehl, G. A., Peterson, T. C., Kunkel, K. E., Gutowski, W. J. Jr., & D. R. Easterling. (2008). *Executive summary in weather and climate extremes in a changing climate. Regions of focus: North America, Hawaii, Caribbean, and U.S. Pacific Islands*. The US Climate Change Science Program. Retrieved from http://metofis.rsmas.miami.edu/~bmapes/SanJuan_NetworkWinter/PAPERS/sap3-3-final-ExecutiveSummary.pdf

Leones J., Colby, B., Cory, D., & Ryan, L. (1997). Measuring the regional economic impacts of streamflow depletions. *Water Resources Research, 33*, 831–838.

Lexow, M. & Edelheim, J. R. (2004). Effects of negative media events on tourists' decisions. In W. Frost, G. Croy, & S. Beeton (Eds.), *International tourism and media conference proceedings*, November 24–26, (pp. 51–60). Melbourne: Tourism Research Unit, Monash University.

Loomis, J. (2008). *The economic contribution of instream flows in Colorado: How angling and rafting use increase with instream flows*. Colorado State University Extension Publication. Retrieved from http://dare.colostate.edu/pubs/edr08-02.pdf

McCoy, J. (2002, August 1). Western drought causing worry about state rafting. *Charleston Daily Mail*. p. C-1.

McKee, T. B., Doesken, N. J., Kleist, J., & Shrier., C. J. (2000). *A history of drought in Colorado lessons learned and what lies ahead*. Retrieved from http://climate.atmos.colostate.edu/pdfs/ahistoryofdrought.pdf

National Oceanic and Atmospheric Administration. (2002). *Drought severity index by division*. Climate Prediction Center. Retrieved from http://www.cpc.ncep.noaa.gov/products/analysis_monitoring/regional_monitoring/palmer/2002/06-29-2002.gif

Natural Resources Conservation Service. (2002a). *Colorado Basin outlook report March 1, 2002*. Retrieved from http://ftp://ftpfc.sc.egov.usda.gov/CO/Snow/fcst/state/monthly/borco302.pdf

Natural Resources Conservation Service. (2002b). *Basin-wide snowpack summary – May 2002*. Retrieved from http://ftp://ftp.wcc.nrcs.usda.gov/data/snow/basin_reports/colorado/wy2002/basnco5.txt

New York Times. (2002, June 16). Fears may be outpacing reality in Colorado fires. *The New York Times*, p. 12.

Nichols, P. D., Murphy M. K., & Kenney D.S. (2001). *Water and growth in Colorado*. Boulder, CO: Natural Resources Law Center, University of Colorado School of Law.

State of Colorado. (2002). *Colorado visitors study, 2001: Final report*. Denver, CO: Colorado Office of Economic Development and International Trade.

US Census Bureau. (2001). *Population change and distribution: 1990–2000*. Washington, DC: US Department of Commerce, Economics and Statistics Administration.

Verlee, M. (2012, July 20). Scorched summers: Tourism takes a hit. *Colorado Public Radio*. Retrieved from http://www.cpr.org/article/Scorched_Summers_Tourism_Takes_a_Hit

Walsh, R., Ericson, R., Arosteguy, D., & Hansen, M. (1980). *An empirical application of a model for estimating the recreation value of instream flow* (OWRT Project No. A-036-COLO). Fort Collins, CO: Colorado Water Resources Research Institute, Colorado State University.

Wilhelmi, V. O., Thomas, S. K. D., & Hayes, J. M. (2005). *Colorado resort communities and the 2002–2003 drought: Impacts and lessons learned* (Quick Response Research Report 174). Boulder, CO: Natural Hazards Research and Applications Information Center, University of Colorado. Retrieved from http://www.colorado.edu/hazards/qr/qr174/qr174.html

Index

Note: Page numbers in **bold** type refer to figures
Page numbers in *italic* type refer to tables
Page numbers followed by 'n' refer to notes